Quest

AN AUTOBIOGRAPHY

LEOPOLD INFELD

CHELSEA PUBLISHING COMPANY
NEW YORK, N.Y., 1980

To My Friends
in Poland

Contents

The Beginning
and the End

I TURNED the ignition key of our car and looked up at the familiar inscription:

Tourist Home. Cosy Rooms with Beauty-Rest Mattresses $1.

I waited for Helen to come out and tell me whether a room were free near a bathroom and whether the landlady looked human.

Without saying it we both understood why Helen went in, leaving me behind the dirty window of our car. This was a New England village. Here I did not care to expose my strange English pronunciation, my unshaven Jewish face, only to be told that all rooms were taken. With Helen it is different. She could have been born in one of the farmhouses sadly flanking U.S. 1, in a world about which I know nothing.

Helen came out smiling. She walked carefully, extending her rounded abdomen, happy and proud of the increasing deformation of her body.

"It's all right. We can stay here tonight."

We entered a room like millions of others. Helen took off her shoes and lay down on the carefully made bed.

For the hundredth time I repeated my question: "How do you feel, Helen?"

"I feel fine; much better than I expected. Bambino did not mind the driving. He only disliked the last two hours. Yes, he definitely disapproved of that."

"And how is he now?"

"I have the impression that he intends to sleep."

"Tell him that tomorrow we shall be in New York and he will get a good rest."

I sat on the armchair looking at Helen and at her shape which grew less and less streamlined with every day. Then I looked at the walls with their cheap prints and ridiculous family pictures until my eyes landed on a wooden cross with its white crucified Christ carved in wood. Helen caught my eye.

"They are Catholics."

"How do you know?"

"These crucifixes are only in Catholic houses."

I went toward the wooden Christ, touched His fragile body on which ribs were distinctly marked, as were the tiny nails with which His hands and legs were fastened. Only the middle of His body was covered with a cloth tied in an unnaturally perfect knot.

For the first time in my life I looked closely at the strange and fascinating wooden face which has seen the tortures of the Inquisition, listened to court proceedings in my country and to the lessons in our schools. He had been present everywhere in my youth. But never before did I have the courage to look straight into those eyes from which now two tiny wooden tears were falling.

With my face turned toward Christ, with my fingers moving up and down along His stiff body, I began to think aloud:

"You know, Helen, when I was a child I was taught that I must not look at Catholic holy pictures. I might go blind, they said. I was scared if I met a Catholic procession on the street. It happened quite often. They went through our streets carrying pictures, big crudely carved figures of the Virgin Mary with the child, and sang loudly in chorus. A Jew, caught in the middle of it, was often beaten if he did not take off his hat. But it was a very great sin for him to take it off."

I turned toward Helen.

"I was taught that our god is the most powerful god in the world, and the only one too. It was bad logic, but I did not see it. If he was the only one, then he had to be the most powerful. To be on the safe side, and to avoid blindness and being beaten, I always ran into the nearest house and waited until the procession had passed."

The subject did not seem to interest Helen. She added a question just to make conversation.

"Anyway, it seems strange. What was a Catholic procession doing in a Jewish ghetto?"

I was glad of the opportunity to talk about an exotic life which she had never seen. I told her about an old Catholic church in the middle of our ghetto. I told how the church was famous because there a daring king killed a bishop during his service; how the king was excommunicated by the pope, how he had to atone by a lonely tragic life for his sins, ending his days in disgrace; how the bishop became a symbol of martyrdom, a saint and a patron of my country. For centuries afterward processions were directed toward this church to remind the faithful that God is above everything and the bishop is above the king. Many other great men lay buried in this church, their dead bodies surrounded by living orthodox Jews.

I remained near the wooden Christ and mechanically touched His small body.

"I wonder whether they would have let me in this tourist house if I were alone without you?"

"Oh yes, they would. They probably imagine that Jews are so terrible that no one who looks half human can be a Jew. And you do look half human, though you need a shave."

I smiled mildly. Our conversation dried up. We sat silently. My thoughts began again to spin around one subject; I felt like keeping them to myself.

We had been driving all day through nearly empty roads from which Labor Day 1939 had swept away the cars. I had promised that I should not try to read newspapers while we drove. Our car formed an isolated system to which the events of the outside world did not penetrate.

I wondered: "What happened in the great world during the last twenty-four hours?"

Helen got up from the bed, leaned over with difficulty to put on her shoes. It was time for our dinner.

As we left the room I said:

"Don't forget to ask the landlady where to buy a newspaper."

"Don't worry, I won't forget. But we shall do it after dinner. Try to relax for a few hours."

We walked along a narrow path, parallel to U.S. 1 alternate. I walked ahead, facing the traffic and courageously protecting my whole family against the possible impact of cars. Among many other things concerning proper American behavior, I had learned from Helen that this is the right way to walk and certainly the one and only way if one's wife is pregnant.

Our landlady had given us good directions. We entered a Dutch diner with a ridiculous windmill outside and commonplace pictures of Dutch girls inside. The food seemed tasteless; the dinner drawn out by my persistent attempts to hide my impatience and by trivial conversation about food and prices. I felt an almost sexual desire to touch a newspaper.

"For three cents I can buy it somewhere and for this price place our isolated system into contact with the outside world. The rest of the world may go to pieces and this New England village will remain as cold and removed from the flow of events as the moon."

Helen looked at my worried face and said:

"Let's try now to buy a newspaper."

I was glad to hear her remark. We again followed directions. We had to go back to our tourist house, then half a mile further and there we should find a drugstore with newspapers.

There it was! Newspapers piled up, vertically and horizontally on the background of yellow racks. I had only to cross the street to find myself again inside the stream of events. Helen cautiously took my arm, looking out for cars, and landed me safely before the drugstore.

NAZIS ENTER CRACOW
FORMER POLISH CAPITAL TAKEN

I did not utter a word. Silently I bought one paper after another. I took out a quarter, laid it on an outstretched hand and smiled idiotically in answer to a remark about the weather. Helen pressed my hand and drew me away from the newspaper stand. I heard her quiet voice and rounded sentences:

"I was afraid of it. I felt that something like this might have happened. You must be prepared for the worst. What I shall say is cruel, but I believe it is the right thing to do. You must be prepared never again to see your family and your friends. Perhaps the situation there is not so tragic as we both imagine. But I am sure that it is better for you to anticipate the worst possible outcome and try to be prepared for it."

She spoke slowly, weighing each sentence, leaving long intervals between them and not expecting any answer. She pressed my arm firmly, then she said:

"Try to think about me, about our child and about our life together. We are happy. I feel guilty that in these terrible times I am still so full of thoughts of our personal life. We must both try to think about our future. We have so much to look forward to."

I knew all that. I knew that in a few weeks I should be back at my university, preparing lectures, doing research, bothering about abstract problems, patiently building—as all scientists do— a bombproof shelter to soften the impact of outside events. My predominant feeling was that of guilt.

"Here I am, safe, free, for the first time living a seemingly sheltered life. A goal for which I fought bitterly through many years seems to be reached. I am in a profession which I believe is the most pleasant and the most exciting a man can have. I chose it in my childhood and achieved it after twenty difficult years of struggle against odds. Yes. I have it now. I escaped from my country. I left my friends, my nearest relatives, not willing to share their fate." I was ashamed of a thought which appeared again and again despite my attempts to suppress it. "Are you not lucky? Imagine yourself in Cracow now."

I tried to cover contempt for myself by being cheap and melodramatic in my spoken thoughts. I said:

"It is terrible. The whole country will be destroyed. I shall never see it again. I shall never be able to show it to you. I feel like leaving everything and going back to suffer and die together with all the people among whom I grew up."

I felt how false and insincere these words sounded. I knew that they expressed only one weak overtone of a passing mood, that this overtone was artificially amplified to cover the underlying sense of relief that thousands of miles divided me from my country. I knew that under the melodramatic overtone there was the strong desire to stay with Helen, to await the birth of our child, to avoid battle fronts and to preserve my life. Thoughts strengthening this basic desire poured into my head. "Poland was beaten before the war began. It was beaten by its own reactionary regime; it was beaten by its own treatment of peasants, workers and Jews."

Helen took, or pretended to take, my melodramatic mood seriously. She argued quietly:

"Only by staying here, by establishing your own life, can you do anything in the future for the people you left behind. You remember that, a few weeks ago, we discussed how to bring your sister's son here and that I always understood your obligation toward your family or friends. A time may come when we shall be able to do something for them. You can't do anything now. But you ought to talk about all your thoughts and troubles as much as possible. I know that all this is very hard on you. Only imagine how much harder it would have been to bear all these events in the world a year ago when we did not know each other."

We went slowly back toward our tourist house, Helen leading me gently along the unreal road. I carried the bundle of papers under my arm, almost physically conscious of their burning contents. The strange setting, designed to last only one night, reduced the whole outside world to the shadowy background of a bad dream.

Back in our room, I deliberately reopened the wounds by reading all the descriptions in all the papers, even when they were identical. I read how the German general visited Pilsudski's grave in the old Polish castle, how Cracow was taken without resistance and how the defeated Polish army was retreating.

I remembered vividly the talks, the speeches and the articles which I had read three years before in Poland. Broken phrases in a harsh, loud, assured tone sounded clearly in my ears: "Our great Polish army will defend our fatherland to the last man; our historical role is always to defend civilization and Christianity." When the liberals reproached the Polish youth for beating Jewish students to death at the university the excuse was: "Youth has its own right. Our youth is patriotic and beats Jews because they are the enemies of our country. The same youth will give the last drop of its blood to defend the Polish soil, and with its strong young arm and burning heart it will annihilate any invader who may attack us. Our splendid army, proudly conscious of its heroic past, is ready for any emergency."

OLD POLISH CAPITAL TAKEN WITHOUT RESISTANCE

I tried to transplant myself to Cracow to elaborate on the picture of the day there, a day so different from that which I had spent driving along the peaceful American highways. Scenes began to form slowly in my mind: simple pictures, abstract and schematic as a mathematical formula, dark and cruel as death. Their background was the street on which I was born, the principal street in Cracow's ghetto, with its gray houses and the uneven pavement covered with dirt, papers and saliva; but this was no longer the old, noisy street full of bearded, gesticulating Jews. In my present picture the street, the houses, the rubbish on the pavement were still there. But the street was empty. As in a ghost city from which every human being has run away, no one walked the pavements. Only the old walls, the old pieces of papers, the old dirt covering the pavement remained. The windows were closed, all of them. And so were all the shops.

Then suddenly a rhythmic noise could be heard. The noise grew. It was the noise of rhythmic steps and of heavy boots beating harshly on the pavement. The boots came nearer; brutal faces and brown uniforms emerging from big black boots became visible: German troops marching through the Jewish ghetto in Cracow. They were passing our house, and all my nerves shivered in the rhythm of the passing steps. Then a harsh command. The command was repeated like a multiple echo in a wave of voices carried through the empty street to the crowded homes. Suddenly the sound of marching boots ceased. The suspense grew. A German officer divided his company into small groups. In perfect order, thoroughly, systematically, they would search the small stuffy apartments. Not one door, not one corner would be missed. They would come to the house where I lived. I heard the heavy steps and the loud banging on the door. The picture began to fade. It became shadowy, removed and unreal. My imagination refused to elaborate further details. It gave me but one more glance at the frightened face of my younger sister as she stretched out her hands hopelessly toward her son. Her husband was not in this picture. He was somewhere with the retreating Polish army, dead or tortured by pictures a thousand times more vivid and gruesome than the shadows which were my creation.

The scenes melted slowly away, but I knew that they would return again and again to my conscious thoughts and unconscious dreams.

"Millions of people killed in China! Thousands of children starve in Madrid! Jews in Poland flogged and executed!" In a few weeks I shall lecture two hours a week on the theory of relativity and drink orange juice for breakfast.

"Millions of people killed in China!" I nearly went to China for a professorship, just before the war with Japan started. Wasn't I lucky?

"Thousands of children starve in Madrid!" Some years ago I was recommended by Einstein for a professorship in Madrid. I nearly went there. In London I had a conference with the Spanish ambassador in which he described vividly to me the beauties of his country. Before the formalities were completed Spain was drowned in blood. I escaped the horrors of besieged Madrid. Again I was lucky.

"Jews in Poland flogged and executed!" Three years have

passed since by a superposition of strange accidents I left Poland temporarily. A year later the way back was closed. I had escaped from the greatest of all horrors.

These associations, binding the fate of nations to my own ego, were as natural and human as cowardice and the instinct of self-preservation. They may be smug and low but they are common to nearly all of us. To save our faces and to clear our consciences we give money to causes, showing our sympathy by putting on sad expressions or by making loud speeches.

I grew up in a world which is being destroyed. Through my whole life I fought the dark, narrow atmosphere of the ghetto in which I was born. Throughout my life I have yearned to see it melted by the progress of passing time. Never did I think that the destruction might be accomplished suddenly by one blow of an armed fist governed by hate.

Never shall I see again the buildings and the shops, the bearded faces and the dirt of the street. Never shall I greet my old friends with a smile which erects a bridge over passing time. Once I knew that my longing for my home town, for the sound of my native language would always be in me but that it would be softened by the knowledge that a return was possible, that I should be able to throw one more glance at the world of my past which I despised and loved.

There is no return now. My pictures will become shadows, their background darker and their features less distinct with every day. I shall never be able to test them with reality and to prolong their endurance. Not long ago the world of my memories was still as real and existent as a postage stamp on a letter which crosses the ocean. This world is now strangled and driven to a sudden death. Rows of soldiers, guns and a stream of blood lie between me and the world of my youth. I must revive the shadows of my memory before they disappear into the dark background, before I begin to doubt that this world ever existed. I must throw one more look upon the image of the path which led me to the present day.

BOOK ONE

The Ghetto

I

Wʜᴇɴ ɪ ᴡᴀꜱ ꜰɪᴠᴇ I was sent to a Jewish school. The room was small, the benches narrow, hard and painfully uncomfortable. We boys sat so near to each other that our arms overlapped and each was pressed by his neighbor. In singing chorus we repeated the letters of the Hebrew alphabet again and again, until we recognized them. Then came the combination of consonants and vowels. We repeated, repeated, repeated mechanically until the letters in the book created the right nervous reaction and the proper sounds came automatically.

The windows of the room are closed. They will remain so for the whole winter; otherwise the cold might come in. The air is heavy and odorous. The smell of potatoes and onions, cooked by the teacher's wife, enters the already foul air of our room from the neighboring kitchen. Our teacher carries a stick and is dressed in the long filthy silk suit worn by orthodox Jews in Cracow. Each of us boys is about five years old. The teacher listens to our singing chorus. His trained ear catches a voice that is off key. Slowly, with his stick raised, he comes nearer to the place from which he can reach me. The suspense grows and I try to sing as loudly and correctly as I can. I am saved. The stick falls heavily on the back of my neighbor. The chorus, guided by the voice of our teacher, repeats with greater intensity our monotonous song. The beaten boy dries his tears with his elbow and the teacher looks for another victim.

Every morning when lessons are over the teacher and his helper take us back in groups to our homes. They collect us again for the afternoon. Each morning and each afternoon we breath together the same air, repeat and repeat until we can read the meaningless words of the holy language in which God converses with the angels. We read words which we do not understand; we read them slowly, making mistakes, then fewer mistakes, until we read them fluently. The first step of our education is finished.

My education was determined by the old rules of the ghetto, a small part of Cracow. It was an island surrounded by a strange world. Outside it were Wawel, the castle of the Polish kings, full of memories of past splendor, wars and love affairs; the simple and impressive Gothic church of St Mary, which with its two towers dominated the town from the old market where peasants sold food and flowers; the Jagiellonian library and its courtyard with curved arches built in the most noble Renaissance style; narrow streets, houses, churches and public buildings which had stood for centuries.

All this was outside the ghetto, outside the district Kazimierz, named so in memory of a great Polish king who preserved peace, had a love affair with a beautiful Jewess and, five centuries ago, opened the doors of his kingdom to the persecuted Jews expelled from Italy, Spain and Germany.

There in Kazimierz, in Cracow's ghetto, thirty thousand Jews, squeezed into a small district, breathed the air in which the smell of poverty was mingled with the poetry of religious ecstasy. The Polish city of Cracow, ruled in the time of my childhood by the Austrian Emperor Franz Joseph I, was far removed from the Jewish ghetto. The distance was much greater than the four thousand miles which now divide me from my home town. To the ghetto Jew the Polish city was a strange and hostile world. The distance could be covered in ten minutes' walk. But the two worlds were separated in thoughts and ideas by the chasm of centuries. They faced each other in mutual ignorance, scorn and hate. The separating wall was built by religious prejudice, language difference and mistrust. Here in this ghetto, at the beginning of the twentieth century, I began my preliminary studies arranged and conducted according to tradition established hundreds of years before. The new century was two years old, as was the quantum theory which marks the beginning of modern physics.

In the old days Jews studied their Bible and their *Talmud*, lived and died in their ghetto and produced many learned rabbis. They did not want and they did not need to know the language of the country. Many of them never saw a Gentile. But even if they had seen one, why should they speak to him? They talked among themselves in Yiddish and they prayed in Hebrew. A Jewish shoemaker in the ghetto made shoes for Jews, a Jewish shopkeeper in the ghetto sold to Jews and bought from Jews. The ghetto formed an isolated system. Its inhabitants studied the great books of Jewish learning and kept aloof from the outside world.

For the Jews in the Cracow ghetto the world was divided into three compartments. One was for Jews. They were held together by their religion, synagogues, dress and language. In the second compartment were the great unknown: the Gentiles. The less one knew about them the better. They had been invented as a punishment for Jewish sins by the God who knows and remembers everything and punishes sons for the sins of their fathers up to the seventh generation. The third compartment was reserved for one man only. He was above the Jews and the Gentiles. He was the Emperor Franz Joseph I, blessed by prayers each month when the rim of the moon first appeared in the heavens. A just old man who liked the Jews, he was neither Gentile nor Jew, but apart from and above everything. It was he to whom God had given the greatest power in the world, and his picture was engraved on each piece of money, earned in hard labor and spent with great reluctance.

The compartments must be kept separate, especially the compartments for the Jewish and Gentile worlds. But for some strange reason the emperor wanted the compartments to be less distinct and decided that all his subjects, and therefore Jews too, must have a four-year secular education in free Austrian schools. The Jews did not like the emperor's decision. The emperor himself spoke the German language, similar to Yiddish. Every Jew can understand German because he understands Yiddish. He can even speak German by speaking a properly spoiled Yiddish. But Cracow public schools taught Polish and not German because Austria left the Polish language in the schools when it grabbed Poland. So the Jews had to learn a foreign language: a difficult, unnecessary task, and perhaps even a sinful one.

In the synagogues everyone prayed in Hebrew. But from time to time one had to interrupt his prayers and say something to his neighbor. The proper language for this was Yiddish. But even German was permissible. Polish, however, would be quite out of place. It was the language of the native Gentiles who were hostile to the Jews. The emperor himself did not know Polish, and now the Jews by his will had to study this strange language.

Thus I was sent to public school when I was six years old. At this time I could already read Hebrew. My parents did what all other parents in the ghetto did. They sent me to the public school in the morning and to the Jewish school in the afternoon.

My parents lived on the periphery of the ghetto, in Kazimierz but near to districts partially Gentile. They felt superior to the inhabitants of the inner ghetto who lived the same life as had their ancestors. My parents approved of mild, slow progress. They drew their religious line in a well-defined way. They were proud both of the progress they had achieved and of the tradition which they preserved. What they really did, like all groups inside the ghetto, was to mix their own religious cocktail, convinced that their mixture was the most tasty to God and man, that any deviation to the left was against the essential religious rules and any deviation to the right uncivilized and superstitious religious exaggeration.

Life forced them to take this stand and to defend it. If the Jewish religion is taken literally, it becomes a sin to breathe without at the same time contemplating the greatness of God. A man who wants to earn his daily bread is forced by life to cut out some portion of the religious prescriptions and will try to find justification for his own deletions. This is what my parents did. My mother was the first in her family who did not cut her hair and wear a wig after marriage. My father was the first in his family to wear European dress. He refused to wear the silk overcoat with a fur collar and the satin cap with thirteen small fur tails which religious Jews in Cracow wore each Saturday.

Once having made a choice of rules for himself, however, he clung to them consistently. His habits, rules of behavior and details of everyday life and dress were all well organized. He always arose early in the morning, went to the synagogue to say his morning prayers, came home for breakfast to eat the same food at the same hour and then went to his shop to work. The shop was just outside the ghetto. The sign read:

Leather and Shoemakers' Supplies

An appropriate colored picture made the idea clear to shoe-makers who could not read. I liked the two dark shop rooms placed in the yard of a big house. I liked to breathe the sharp smell of leather and to look into small compartments where all the different shoemakers' accessories lay in perfect order. As a young boy I liked to watch my father take a paper pattern of a sole and trace it on the leather. Then with a curved knife he would cut the leather carefully on the drawn lines, his face red from exertion. He would weigh the piece, calculate the price and write it down with chalk on the leather, calling it out loudly in Yiddish or very bad Polish, depending on the nationality of the customer.

My father had inherited the leather business from his father and had decided that I should carry on the tradition. I was to do what he did, but perhaps better and on a larger scale. I was to be a religious Jew as he was. The religious line which he drew for himself was also for me. As he had rebelled against religious dress for men and wigs for women, I might also. But somewhere near where he had stopped I must stop. I was allowed to draw the line just a little to the left of his. Though he wore a cap all day (it is a sin not to cover the head before the eyes of God), he did not mind my wearing a cap only when eating and praying. He would not have objected to my moving the leather shop a little farther outside the ghetto. But beyond such small changes he was unwilling to let me go. He expected me to say my prayers every day, to eat kosher food all my life and to continue his work in the business. A business was something eternal, something which ought to pass from generation to generation, and it was my duty as his only son to preserve it.

My mother was short and fat, with an extended abdomen which was forcefully pressed into a tremendous corset when she went outside our home. She had a smooth, shining complexion and a sweet smile. Her philosophy of life was simple and could be put in a few sentences:

"The most important thing is to trust in God."

"Everything is fate. I believe in fate."

"A good son must always listen to his father and mother."

"If health is good then everything is good."

Indeed, the health of her children was her greatest concern. Sometimes God, sometimes faith, sometimes obedience to parents, was the most important thing in the world, but chiefly it was health. Her good nature was undisturbed by problems which she solved in a sweet, smiling way by transferring them to God who best knew how to deal with them.

"Why do people get whiskers when they grow old?" I asked my mother.

"Because of nature," she replied.

"What is nature?"

"Nature is God."

I could not ask further. After some training a child considers God an elementary concept at which all questioning stops. Parents make good use of God; He simplifies the task of explanation.

Both my parents belonged to the upper class in the ghetto. Our house was clean, well kept and dignified by the presence of an M.D. In our apartment were three rooms and a kitchen. But we spent all the hours of the day in only one of them. It contained two iron beds painted with red flowers, a sofa and a table on which we ate our meals. I slept on the sofa; my two sisters slept together in one bed and my maternal grandfather slept in the other bed. This arrangement lasted until I was eighteen, practically as long as I remained at home. Here in this one room we children lived, ate, slept, read, dressed and undressed, prepared our lessons and prayed. My parents had a small separate room to which they retired for the night only. The third room was pompously called a "salon," kept under key the whole week and opened only on Saturdays when guests came to visit us. Our guests—chiefly members of the family—knocked at our door, since ringing the bell on Saturday was sinful, and were proudly admitted to the salon. They brought their children to play with us. The women talked about servants, children and health; the men about synagogues and politics, impatiently waiting for the evening which allowed them to light the lights and to break the Sabbath by smoking cigarettes, forbidden during the day on which God and man must rest. Only on Saturdays did our family expose its salon full of china, family albums and heavy furniture.

During the week I saw my parents very little. They returned from business about nine in the evening, and these evenings and Saturdays are the only memories I have of childhood contacts with my parents.

My grandfather came to us when his wife died. The death of my grandmother and my mother's cry in the middle of the night are the earliest memories of my childhood. My grandfather was a tough old man who stayed around the flat and shop the whole day. His only activity was to mumble his prayers quickly and mechanically. He muttered to himself, smoked the cheapest and most ill-smelling cigars, hated everyone and in turn was disliked by everyone. His religion was a collection of rules senselessly accepted and mechanically performed as unpleasant duties. He wanted to dominate everyone. Since it was difficult and dangerous to attempt it with my father, who kept him, he exerted the full force of his pressure on us children. I was a special object of his hate. When I played too noisily, interrupting his deep contemplations, he lifted me by the collar, pushed and beat me and threw me violently outside our flat, closing the door behind me. I waited on the stairs, crying and pleading, but nothing moved him. He would let me in only when the time of my parents' arrival was near.

My grandfather had a habit of saving his evening meal to be eaten later. He kept his plate on the top of the wardrobe. In the middle of the night he would awaken, light a candle, take out his shirt, catch fleas for a while and then begin to walk around the room, noisily eating his food with his toothless mouth. He never uttered a kind, tender word. Day and night, for months, he wore the same dirty, striped, long flannel underwear. I still have a clear picture of the striped underclothes, long white beard, sharp eyes and lighted candle, haunting our home day and night. He died a long time ago, but I shall never be able to say anything kind about him. He symbolized for me what can become of religion when the skeleton of rules is observed but no meaning attached. Religion did not soften the sharp edges of his cruelty; on the contrary, it gave him one reason more for imposing cruelty on others through intolerance. My grandfather would have enjoyed being an Inquisitor if he had had the opportunity. Instead he visited his spitefulness upon the lives of us children. He was always present, hating, fussing, humiliating me and my friends whenever they came to visit me. Even much later, when I was a grown boy, he would get up from his bed in the afternoon in his striped underwear and drive my friends from the house.

Two thoughts, two desires, recurred in my childish mind. I thought that life would be quite different if my grandfather would die and all the Jewish schools in Cracow be burned. Neither of these events occurred as long as I was at home.

My mother dismissed all my complaints with the same remark: "He is your grandfather and he is my father. At heart he is a good man; he is only very nervous." Her reasoning had the virtue of early scientific explanation; it stated qualities so well hidden that their presence could never be disproved.

I also tried to restrict my studies in the Jewish school, but my father's response was: "I don't want my son to be uncouth." Sometimes the reaction was stronger: "You will grow up like a *goy*." (*Goy* = Gentile + a condescending tone.) The possibility that I might be baptized and become a Gentile was the strongest threat ever used against me in my childhood. It was associated in my mind with a picture of a lonely man walking through the ghetto whom everyone had the right to stone and spit upon.

My parents expressed ideas to which they were exposed in their childhood, trying to believe and to make their children believe that we were exceptionally fortunate to be born Jews and that we should be proud of it. It was this belief which kept the ghetto intact against outside pressure. Only by nursing a superiority complex could the misery of ghetto life, with its poverty and lack of opportunity, be borne. How is it possible to stay in a ghetto, to be kept far from the outside world, to cultivate the same kind of life for generations without escape through religious ecstasy and without the belief in a God whose will is obeyed and who chose the Jews to suffer and be redeemed by the Messiah?

The division of the whole world into Jews and Gentiles was one of the first things I learned in my childhood. Who were the first Gentiles with whom I came in contact? I owe them much, indeed much more than I realized at the time. I owe them my knowledge of the Polish language, devoid of Yiddish pronunciation. Without this seemingly trivial emancipation I should have found the doors leading outside the ghetto still more hermetically sealed.

When I was born my mother followed the fashion of the time and hired a Gentile wet nurse for me, a symbol that some modest wealth had been achieved by my family. I remember, for years afterward, a fat woman coming every few Sundays to see me, bringing with her candy and a loving smile. My parents told me how wonderful she had always been to me. Years later, when she was ill and dying, she sent her son to fetch me. I was at that time far outside Cracow. One of my sisters went to her, arriving just in time to see her still alive. Later my sister told me how my old wet nurse lost consciousness, talking tenderly and incoherently about me, and how the words, "I hear his steps, he is coming to see me," and my name repeated again and again were the words with which she died.

Our servants formed my next connection with the Gentile world. They changed frequently, but some of them stayed for years, others went away and came back, having found our place better than others. The servants in the middle-class families had a hard life. With no restrictions on working hours and servants' rooms unknown, they worked like slaves and were allowed to go out only on Sunday afternoons. Nearly every one of them had an Austrian soldier sweetheart, the son of a Polish peasant, conscripted for three years' military service in the emperor's army. Sometimes our servant's soldier came to the kitchen for a brief visit. He took little notice of my presence, and I could remain in the warm kitchen. I can still see the maid's full breast overflowing through the soldier's fingers, a picture which haunted me for a long time with its aggravating strangeness.

The servants were mostly peasants' daughters pushed by the poverty of the Polish peasants toward the town, to meet a life of hard work in which religion and sex were the only possible outlets. All the maids' love stories were similarly tragic. Always the same pattern: soldier, pregnancy, increased earnings as a wet nurse, her own child sent to the country to a woman called "manufacturer of angels." The child usually died from undernourishment and mistreatment, and the old story began again, the cycle finally interrupted by marriage or a whorehouse.

From these poor peasant girls I learned a Polish which was at least free from Yiddish pronunciation. But I owe them more than this. I remember the sympathy which some of them showed me when I was humiliated by my grandfather and their kindness and gentleness toward me and my sisters.

All the Gentiles I met in the first years of my life came from more or less the same social level. There were the Polish shoemakers who bought leather from my father, the Gentile janitors and their wives—usually ex-servants—garbage and repair men. The belief in the inferiority of the Gentiles which I learned at home was strengthened by the first experiences of my childhood. The Gentile world with which I came in contact was—

accepting the point of view of my surroundings—socially inferior to ours. The only impression which I could have formed as a child was: "The Gentiles are a collection of good and harmless people, supported by the socially superior Jews."

The whole picture changed rapidly and left me badly confused when at the age of six I started my school education. My mother took me to the elementary school very near to our home. It was a yellow building with a tower and a clock, which had previously been the old city hall. We passed the schoolyard with its iron fence, went through a wide corridor full of brightly colored pictures to the principal's office. The principal asked me my name in a kind voice and took me into a big classroom full of light with a colored picture of the emperor hanging on the wall. The teacher indicated a comfortable seat with a place to put my exercise book, a place all to myself, where I did not need to push or to be pushed by anybody. The first rule of behavior which I had to learn was to be silent and not to talk with my neighbors. Often I heard the teachers remark:

"Quiet, this is not a Jewish school, this is a Polish school," or, "You seem to think you are in your Jewish school." I found the school wonderful, and I loved it from the beginning. Everything was easy, interesting, exciting and pleasant.

The boys represented the social structure of our neighborhood. Among forty boys there were only one or two Gentiles, the sons of janitors in Jewish houses. They were ragged and undernourished and did not do well in school. Contact with my fellow students could only strengthen my belief in the Jewish social and mental superiority. But there was something here which smashed all my belief in this superiority. The school was for me a temple detached from life and the teachers superior human beings who knew everything and to whom God had given the power to control five hours of our daily life. But among the teachers were both Jews and Gentiles. Slowly I learned their names, and when I told my mother how wonderful one of my teachers was she remarked:

"He must be a nice Gentile."

So I began to divide the teachers into Jews and Gentiles. I

became completely confused. I discovered that among the most powerful and best human beings, among the teachers of my school, were not only Jews but also Gentiles. But more than this. The Gentile teachers were kind and at least as good as the Jewish. Mostly from them I learned the touching drama of Polish history. I learned how Poland, once great and free, lost her independence. I learned about the Polish kings, about Cracow, my native city, the capital of the great kingdom, about Kazimierz the Great who admitted the persecuted Jews to Poland and how they developed the commerce and wealth of the country.

The school slowly formed in my mind an unreal picture of the outside world with its tragedies and struggles, all so different from the world in which I lived.

All my teachers, Jews and Gentiles alike, were Polish patriots. The Polish patriotism of my teachers was not in contradiction to their loyalty to Austria. Their argument, gradually conveyed to us children, sounded something like this:

"Poland was a great and independent country. It lost its independence partially through its own faults, through its own political decay, but mostly through the selfishness and greed of other countries. Now Poland is divided among Russia, Germany and Austria. Russia and Germany, two greedy and cruel nations, grabbed the greatest part of the country, repressing the Polish language and Polish thought. Austria is the only country which allowed the Poles to cultivate their language and science in two Polish universities. We must therefore be loyal to Austria and to the Emperor Franz Joseph I, because he allowed us to develop our great inheritance. We must keep it and continue its growth until the day when a united and independent Poland will be reborn."

The belief in Poland's resurrection was nursed and strengthened by the great Polish romantic poetry of the nineteenth century. Poland was compared to Christ, having to die and to suffer for justice, peace and the future happiness of other nations. As Christ suffered and died to bring love and brotherhood to the human race, so Poland, the Christ among nations, took the burden of others' sins on its shoulders. But as Christ rose from the grave alive, so also would Poland again be free, carrying the light of justice and love as a vivid example of brotherhood to all mankind. This messianic idea of Poland's historic mission may sound naïve. But it captured the imagination of the youth and kept alive their belief in the future. And, above all, this faith was expressed in Polish poetry with power and in a form which makes me believe even today that Polish poetry is one of the greatest in world literature.

This was the atmosphere which we breathed in school. Poland was, at this time, a suppressed country, and nothing is so easy as to awaken in young children the feeling of sympathy, the desire to work for the future of the suppressed.

In the afternoon I had to study the Bible in a more advanced Jewish school. My father brought me to the schoolmaster, a man with a severe face, a long gray beard and an extended belly. In his hand he carried a short cane. He looked me over gravely in front of his assembled class, then opened a Hebrew book and let me read. Satisfied with my knowledge, he took me to his helper from whom I was to obtain some individual tutoring before I could join the advanced class. This was the unbreakable rule.

The helper was a young man with a small black beard and pock-marked face. He put the holy book of the Bible on the table, opened it at a worn-out and dirty page, placed a wooden pointer with a sharp end above a certain Hebrew word and commanded:

"Read this!"

Slowly I read the word.

"Repeat!" I repeated. "Repeat!" I repeated. So it went five times. Then he said a German-Yiddish word, followed by the same command: "Repeat." I repeated obediently. After five repetitions I knew both words by heart, not understanding either of them. But I got the idea. These were two words in two different languages. They didn't convey any meaning to me, but I knew that they were bound together by the magic equivalence relation. His pointer jumped to the next word. The method began to make sense. Soon, after a few repetitions, a new German-Yiddish word would be thrown to me. I would have to catch it quickly like a ball and repeat it again five times. The wooden pointer moved slowly along the line, then down to the next and the next. In his other hand the teacher carried a stick. But he saw from the flow of the lesson that the stick might be unnecessary. So he put it carefully on the table, and having freed his one hand, he started to pick his nose, slowly, thoroughly and with dignity.

Only years later I found out what I was supposed to have been doing. I was learning by heart and translating into obscure words the following passage from the Bible:

And the Lord called unto Moses, and spake unto him out of the tabernacle of the congregation, saying,

Speak unto the children of Israel, and say unto them, If any man of you bring an offering unto the Lord, ye shall bring your offering of the cattle, even of the herd, and of the flock.

If this offering be a burnt sacrifice of the herd, let him offer a male without blemish: he shall offer it of his own voluntary will at the door of the tabernacle of the congregation before the Lord.

And he shall put his hand upon the head of the burnt offering; and it shall be accepted for him to make atonement for him.

And he shall kill the bullock before the Lord: and the priests, Aaron's sons, shall bring the blood, and sprinkle the blood round . . .

Why did my education of the Bible start with these words? Because this old custom must be obeyed by all children of Israel, and Jews have for it, as for everything else, a rational explanation. It is even very simple. There obviously cannot be a better and more innocent deed than that of bringing offerings to the Almighty God. Nor can there be anything more innocent than young unspoiled children who have not had an opportunity to commit sins. Then offerings are innocent and children are innocent. Is it not right that the two should meet as soon as possible?

After a few tutoring lessons I joined a class conducted by the gray proprietor, to sing in chorus weekly portions of the Bible. For hours each day we all sat in the small, dirty room, two or three of us squeezed together over one Bible, and we repeated in chorus a word with its German-Yiddish translation, a word and its translation. Each Sunday we started a new section. It went unevenly and it seemed difficult. But we repeated the same sentences on Mondays, Tuesdays . . . and with each day it went more smoothly, until on Friday afternoon we could recite the whole story in a quick, well-organized chorus.

The next week we started the next installment. Day after day, week after week, I spent afternoons in the Jewish school plunged in a hopeless ocean of boredom. I turned my eyes constantly toward the clock hanging on the wall, the slowest clock in the world.

The routine of afternoon teaching was interrupted by a half hour interval of rest for teacher and children. We stayed at our places; we could not go outside because there was no outside. The school was a small, poor flat in a poor ghetto house. So we sat down and talked. Whenever my mother gave me a cent I would buy two pieces of candy, one of which I tried to exchange for a string, a piece of metal or a postage stamp. During this half hour our room was a stock exchange, full of talks about our wonderful fathers and mothers and serious quarrels over which of us had better and richer parents.

As we grew older our talks became less childish. Once my neighbor whispered to me:

"Do you know what a whorehouse is?"

"Tell me."

"It is a house where men pay for entrance, and then they can see naked women and touch them."

I was skeptical and did not believe in the existence of such houses. But my cheeks began to burn. The boy said to me:

"When we go out I will show you such a house. It is not far from here."

We went to see the whorehouse; it was practically on my way home. I had to believe in the boy's story because there was something strange about the house. The green curtains were pulled down and the house was silent and isolated, an island in the middle of the ghetto. From then on I passed the house each day in the hope that I would see something happen. Only once did I see a man look cautiously behind him and then quickly go inside. A picture began to form in my mind, fantastic, repulsive and at the same time attractive; I imagined fat women with big breasts and extended stomachs, the most common type in the ghetto, walking in long white nightdresses such as my mother wore, with men running after them, lifting their skirts and touching their bodies. It was as fantastic and sinful as the picture of hell.

The Bible also did its share to stir our imaginations. After months of repeating meaningless words we began slowly to understand something of what we read. Gradually we recognized the same words and the same translations appearing in different connections and by this involved method guessed the meaning of words and sentences. The words "pregnancy" and "conceive," which appeared in the Bible, were associated with vivid pictures readily provided by our childish imagination.

Passages like:

Come, let us make our father drink wine, and we will lie with him, that we may preserve seed of our father.
And they made their father drink wine that night: and the first-born went in, and lay with her father; and he perceived not when she lay down, nor when she arose. . . . Thus were both the daughters of Lot with child by their father.

remained in my memory until I could guess their full exciting meaning.

About six o'clock the Jewish school was over and I was allowed to go home to prepare lessons for the next morning. Often I had bad luck. My sisters were out, the servant shopping and my grandfather the only person in the flat. Impatiently I would ring the bell. The door would open a crack, my grandfather's eye look through the narrow opening, and then a hand would quickly slam the door in my face. I employed different tactics, depending on my humor. Sometimes I sat on the stairs burning with hate while I waited for our servant.

"How dare he treat me this way? He is a very bad man. I am the most unhappy one in the whole family. Why are not my sisters here to help me? They have a good life. They don't need to go to the Jewish school. They have the whole afternoon free and they may do what they like. Even our bad grandfather treats them much better than me. I am the one who is worst off."

Those thoughts were in vivid contradiction to the words which I repeated daily in my morning prayers:

"Blessed be the Almighty God who created me a man."

I consoled myself with thoughts of the future in which I should be strong and important, my grandfather afraid of me and regretting all he had done.

But more often I took fate less philosophically and began a war of nerves. I rang the bell constantly. Then my grandfather would suddenly open the door and try to hit me with his umbrella. At this I would run away quickly and start the game over again until the servant came.

Once I ran away too slowly and the blow of the black umbrella was especially painful. I went crying into my parents' shop. My father was angry and spoke to my mother in rapid Yiddish, believing I would not understand:

"He has no right to do that to my son. Only you and I have a right to beat him if he behaves badly."

"Go and tell him that," my mother said, "he is afraid of you."

"Do you want me to leave the shop and go to your father? He is your father, not mine. I will get angry and say too much. Tell him he must leave the boy alone."

With a feeling of triumph I went home with my mother. I anticipated a scene and imagined my grandfather cowed, humbly promising never to persecute me again. My mother, red and nervous, opened the door with her key, took off her coat, walked past my grandfather and went straight to the kitchen.

"Where have you been from six to half-past six?"

The servant humbly said that she had gone shopping.

My mother was angry. "That takes only five minutes, not half an hour. You met your soldier and talked with him for hours. If this happens again, you may leave our house."

Then, having vented her first anger, she turned toward Grandfather.

"Why did you not let the boy in?"

My grandfather answered gruffly:

"I am not a janitor in your home. You have a servant. You may bully her but not me."

My mother hesitated for a moment, then collected all her courage for one sentence:

"It would not cost you more to open the door to the boy."

My grandfather howled furiously from his toothless mouth:

"I would have opened to him if he were a nice boy. But he is a disgrace, not a nice boy. He rang and rang and nearly made me deaf. If he were my son I would have broken all his bones into small pieces."

His anger increased as the speech proceeded, ending in incoherent oaths. My mother repeated more to herself than to her father:

"It would not cost you more to open the door to the boy."

The weekday routine was interrupted by Saturday. In preparation for this important day my mother washed my face and neck after dinner on Friday night. Every few weeks, on Friday, my father took me to a public bath where we bathed together in one tub and he washed me thoroughly. With the tidiness characteristic of everything he did, he prepared a new suit for me, fresh shirt, socks and shoes each Saturday morning.

Before giving me the new outfit he carefully examined my hands, neck and ears, making the last necessary corrections. Saturday made him especially severe, as though it were my fault that smoking was not allowed on this day. The slightest provocation caused an explosion. I dreaded the Saturday rows, but all my efforts to avoid them were fruitless. Later I incorporated them into the Saturday program as an inevitable part, like the chopped onion with egg and the roast duck of Saturday's dinner.

"What will happen to you," thundered my father, "if you cannot keep your things carefully? Last Saturday I gave you a new collar button. Where is it now? All right! If you haven't got a collar button you will go without a collar and look like a beggar's son. The next time I will let you go without a collar, and if people ask me in the synagogue why my son looks like a beggar, I shall tell them the whole truth. Must I always be ashamed of you when I take you to the synagogue?"

His face grew red, his gesticulation more violent. He touched his face with his hands, then clapped them together with a loud noise, finally shaking his index finger at my nose.

"If I give you a new collar button it's the last one I'll ever give you. See this collar button?" Then he waved the collar button so near to my eyes that they closed automatically. "I have had it for twelve years since my marriage, and I will have it another twelve years if you will bring it back to me."

The synagogue was inside the ghetto in a narrow, poor street. It consisted of one room in a big old house full of tenants. The house, with a tremendous yard in which a block of buildings could have been erected, contained an interesting cross section of the poorer part of the ghetto. Without plumbing, old and badly kept, occupied by Jewish poverty, its air was sour with the smell of garbage and urine.

With my father I went through the great yard; then we turned left to a narrow hall with a sink and faucet from which water fell in intermittent drops. From the hall we came straight into a noisy room full of dense air and the smell of fur, with high benches and a cabinet where the holy handwritten Torahs were kept behind a richly decorated curtain.

My father was always received with great reverence. He was on the Committee of Eight, which administered the synagogue, and had a seat of honor in the first pew. My father's fine reception was extended to me, and I received warm handshakes and snuff. Like everyone else my father put over his shoulders a large prayer shawl embroidered with Hebrew inscriptions and fringed with white tassels. First he kissed the cloth in the prescribed places, especially the tassels, and then began his prayers.

The prayers were guided by the cantor. He sang the first words of each chapter, and the whole synagogue was his chorus. It was his business to choose melodies, to give the tempo and the proper swing.

Great variations were possible in the ritual. Without thinking about the meaning of the words all of us in the synagogue were actors in a play rehearsed each week. For a few minutes we sat down quietly with our prayerbooks open, murmuring privately to ourselves. Then we all stood up and each of us cried as loudly as possible. Especially the sentence:

"Hear, O Israel, the Lord is our God, the Lord is one," created great enthusiasm. Some of us began to raise both our fists to convince our God that we were his fighting army. Since the emphasis of the sentence lay on the word "one," the proper thing was to draw this word out as long as possible, to pronounce it crescendo to the last letter and then to finish with a deep sigh reflecting all the sorrow of Jewish life. Then again there was a change. Everyone stood up, pulled the shawl from his shoulders over his head, whispered his prayers very quietly, swinging his head and his back in silence and sighing. At certain places in the prayers everyone moved back three steps, then three steps forward, and then beat his breast with his fist, made a deep bow and kissed the tassels. After the quietest part of the performance was finished the cantor began to sing, and the whole congregation joined him in a well-defined way prescribed by inflexible rules.

At first I found the ritual interesting and exciting. I was told to do what everyone did, and I played my part clumsily. But after many weeks I learned the smallest details of the performance and I became more and more bored by the whole procedure.

The prayers were interrupted by reading the appropriate installment of the Torah, for which I had been prepared by the week's study. Ironically enough, by a tradition which seemed designed to make everything meaningless and nonsensical, we children were allowed to leave the synagogue just during the part of the service for which we were so painfully prepared. We usually made use of this freedom to play in the yard and watch its intensive life and to urinate in a small room containing only a hole in the floor, where the strong smell of stale urine forced us to hold our breath, making the act exciting and important.

The service took about three hours, after which my father carefully folded his prayer shawl, put his prayerbook inside, removed his small prayer hat and very quickly replaced it with his top hat so that the sin of standing before God with a bare head was made infinitesimal.

Slowly with the other men, stopping every few steps, talking and gesticulating to make arguments more vivid, we walked through the ghetto streets. The streets had their Sabbath too. All shops were closed. The men and women stepped differently on the pavements during the Sabbath: slowly and with dignity. Men wore their furs and their peculiar caps with thirteen fur tails. Women, children, everyone dressed differently and looked different. The whole ghetto radiated on the Sabbath the reflected light of God's glory.

On Saturday every Jew is a king and every Jewess a queen. My mother waited for us at home with her pearls around her neck and diamonds in her ears. My father kissed her tenderly. By this time he had recovered from his cigarette hunger, and everyone tried to be angelic during dinner. Even if father blew up, he merely described how terribly angry he would have been if it were not the Sabbath. The time reserved for quarrels was definitely before and not after the synagogue service.

But even Saturday afternoon was not completely free. I had to go to the Jewish school for two hours. Not to study the Bible: this was reserved for weekdays. The teaching on Saturdays was taken very easily. God's peaceful spirit penetrated the school; our teacher did not carry a stick because religion forbids it and beating on Saturday afternoon would be quite out of place. The aim of our Sabbath session was to read and translate in the usual fashion the great sentences of our fathers, word for word.

Wise men lay up knowledge.
The thoughts of the righteous are right: but the counsels of the wicked are deceit.

Sometimes a sentence sprang up which was designed to kill our rebellion against our fathers' strong hands:

Withhold not corrections from the child: for if thou beatest him with the rod, he shall not die. Thou shalt beat him with the rod and shalt deliver his soul from hell.

Once more on a peaceful Saturday afternoon we were re-minded that the circle formed around our souls was unbreak-able. We were inside and must remain inside for the rest of our lives.

II

WHEN I WAS TEN YEARS OLD I had my first great battle with my parents, a fight in which my whole future was at stake.

I was just finishing the four years of public school prescribed by the Emperor Franz Joseph I. In the old Austrian school sys-tem parents had to decide their son's future when he was ten. At this age a student who finished public school could start one of two quite different branches of education, different with respect to the social standing of the pupils, to their educational level and, above all, to their future.

The first branch, representing the far more distinguished edu-cation, was that of the *gymnasium*. A boy entering the gym-nasium was dressed in a blue uniform similar to that of an officer in the Austrian army. On his collar he wore stripes in-dicating clearly to which of the eight gymnasium classes he belonged. He was drilled in Latin for the full eight years and in Greek for six. Professionals, army officers and high governmen-tal officials were supplied by the gymnasia. The goal of the gymnasium studies was the *matura*, a severe examination which, if passed, opened the doors of the university to the student when he was eighteen. Then four or five more years of study for the degree which made life so different from that of my parents.

Not many Jewish boys entered the gymnasia at this time. Most of those who did came from an environment more ad-vanced than mine, from parents who had moved outside the ghetto and who had made the first step toward escape and were encouraging the children to take the next. A few came from the ghetto, from parents who, while maintaining tradition, understood the importance of education for their children. I was somewhere on the border line.

It was the year 1908, eight years after the quantum theory was formulated and three years after Einstein's first paper on relativity theory appeared; two events which were to play an important role in my life

I thought the gymnasium would be the door to the outside world, the fulfillment of my desires to learn and to escape from my environment: desires as old as the earliest memories of my childhood. I never doubted that I should be able to pass the entrance examination.

I saw myself dressed in the splendid blue uniform, wearing the cap with "G" on it, with silver stripes increasing in number each year until, after reaching four, they changed to gold and increased again. I saw myself in a class with the best pupils of my public school and all other schools, in one of the seven splendid buildings in which the gymnasia were housed. Since the preparations of lessons for the gymnasium took so much time, gymnasium students didn't attend the Jewish school. So at the same time that I gained my blue uniform I expected to be free of the afternoon misery. The prospect seemed good.

The only other possibility, which I refused even to consider, was to go for three or four years to a "citizen's" school. The citizen's school was attended by children of poorer parents, who did not have the means to support their sons through many years of study, and by those who could not pass the entrance examinations to the gymnasium. The gymnasium was for the socially privileged and intellectually superior boys, whereas the citizen's school represented a lower level. If in his tenth year a boy was not admitted to the gymnasium, his fate was practically sealed, since the matura was difficult and the knowledge of Latin and Greek could hardly be self-taught.

I first brought up the subject with my father as a matter of course. I did it at the only suitable time: Friday evening, when my father was relaxed and in a holiday spirit. It was after dinner. My father drank his second cup of tea, loudly sucking a piece of sugar and at the same time glancing through his newspaper. The whole family was assembled around the table when I started my little speech.

"Next year, when I go to the gymnasium in the morning, I can not go to the Bible school in the afternoon. The gymnasium is very difficult, and I will not have time in the afternoon. My friends who will go to the gymnasium told me so. They will only have a teacher for the Torah, who will come to them three times a week, and I ought to do the same."

My father stopped reading the newspaper and looked grave. My mother became red and frightened. My sisters were interested. Grandfather murmured to himself: "Now he is starting something new. Even on Friday he cannot sit still."

My father, in a voice which excluded the possibility of further discussion, said:

"There is plenty of time to decide it. Who told you that I shall send you to the gymnasium? A good son leaves everything to his parents. They know much better how to decide things. You sleep in peace and let me worry about your troubles."

A week later my father came back to the subject and his verdict was "no." He had thought out the reasons and tried to convince me, starting in a quiet, friendly way.

"I don't want you to be a doctor or a lawyer. The gymnasia are overcrowded. Before you finish there will be so many doctors and lawyers that you will starve from hunger. I want you to be a businessman like me, but better educated. I want you to be an intelligent businessman."

I did not believe that the verdict was final and tried to use my childish logic.

"If I go to the citizen's school I cannot even be an intelligent businessman. I am ashamed to go to the citizen's school. Only the poorest and worst boys go there. I will be the only one from a good family. No one of my friends will be there."

"Then you will be the best boy in the citizen's school. Somebody must be the best; why not you? Better than to be one of the worst in the gymnasium. They will teach you in the gymnasium to be ashamed of your father and mother and to be ashamed to go to the Jewish school. I don't want my son to be ashamed of his parents and to be ashamed that he is a Jew. Much better to go to the citizen's school and later to a good business school and to be an intelligent businessman."

I tried to fight again by emotion rather than logic. My voice trembled:

"Send me to the gymnasium, Father, and I promise you that after four years I will leave it and go to a business school. From the gymnasium I can go anywhere, and I will learn in four years in the gymnasium much more than in four years of a citizen's school. Send me only for four years to a gymnasium and I swear to you that I shall not ask you to send me any longer."

My father seemed sorry for me.

"Now you say that you will go for four years to the gymnasium and I know that you believe it. But then you will say, 'Let me stay one year longer,' then again, 'Let me stay one year longer; it is so near to the matura.' Then the same story with the university, and I shall lose my son. You will see later, when you grow older, that what we do and wish is only for your good. Now you are young and stupid, but you will get your brain later and you will be thankful to us that we did not allow you to go to a gymnasium."

I tried my last weapon: weeping loudly, I repeated over and over again the same arguments. But nothing helped. My father was sure that he was doing the right thing. The battle was lost. I could not bear to see my old friends in their new blue uniforms, and I lost all my old connections. I felt that I had been socially degraded. Slowly I realized that what my parents had done was to close to me the easiest exit from the ghetto.

Depressed and broken, I went to the citizen's school. And here, in a level considerably lowered by the absence of the best boys, I was a very average pupil. I had lost my old interest in school. I could hardly follow the teaching, even though reduced to the low level of the pupils. This fact, appearing clearly in the reports brought home, was used by my father as an argument against me:

"You are not a good pupil now. I am very much disappointed in you. And you wanted to go to the gymnasium! How could you do well in a gymnasium, where it is so much more difficult? You see now that I was right. Parents are always right. They are older and have more experience. In the gymnasium you would have to repeat the same class for three years until they would throw you out."

Whenever I was asked what school I attended I blushed, felt ashamed and stuttered the name of my citizen's school. I avoided my old friends. When I was thirteen I was as deep in the atmosphere of the ghetto as in the first years of my childhood.

III

I OBEYED MY MOTHER AND FATHER and went to a citizen's school and then to a business school to prepare myself for the future occupation of selling leather and shoemakers' accessories. I tried hard to convince myself that I ought to accept my fate, that my parents offered me the fruits of their lives: a good business name and two rooms full of leather.

The formalities of my morning prayers increased as they do for every Jewish boy after he is thirteen. Two leather cubes had to be placed, one on the forehead, the other on the left arm. They contained Hebrew inscriptions which I kept near to my brain and near to my heart so that they would gradually diffuse into my mind and heart through the thick leather covers. The cubes were bound to my head and arm by means of leather strips. Leather again! The same sharp smell as in my father's shop. By leather straps and by the words of God I was bound to my future profession. My morning prayers were a condensed symbol of the life which was ahead of me.

Just when everything seemed to be settled the shape and colors of my world changed. I outgrew the Jewish afternoon school. It ceased to exist for me. I had time for leisure, thought and books. But the essential impulse came not from outside, not from my family, not from my friends. Nature changed the course of my life, and the intervention started distinctly from a strange source: from sex.

More and more often I found myself thinking about exquisite women dressed and looking like the heroines on the stage of Cracow's theater. I imagined scenes in which I touched their bodies, embraced them firmly and was loved with warmth and tenderness. These feelings were so strong that breathing became a conscious act full of pain and joy. The world glittered with mystery and hidden treasures which must become mine. New streams of life poured into my veins. I felt like conquering the world, dreamed about my future fame while I learned book-keeping or lay in bed in the same room with my grandfather.

I discovered books, all kinds of books. Fiction, pornography, popular books about science, socialism, the theory of evolution, physics.

I discovered friends, most of them older than I. We had long discussions about saving mankind, what it means to be a genius and whom we should call intelligent.

Religion became one of the frequent topics of our conversation. We all saw it was silly and meaningless to refrain from writing on Saturdays, to knock at the door instead of ringing the bell, to put the cubes on our heads and arms each day. But most of us claimed that we ought to do these things to please our parents, to avoid lies and deceit.

Once I was lying in bed sick and therefore, in God's eyes and my father's, excused from going through the formalities of putting the cubes on my head and arm. But in God's eyes and my father's I was not sick enough to omit the straightfor-ward prayers from my prayerbook. I got up from bed with a headache, took a prayerbook and began to recite the prescribed daily portion half aloud. But the procedure in which I was so well trained that I was hardly conscious of it began suddenly to be boring and troublesome. Could it be possible that God cared about my mechanical prayers? As a child I had pictured God as an enormous Jew with a tremendously long silver beard, watching everything carefully from above the clouds. But after this naïve and childish picture vanished nothing took its place. The prayers were a duty which I performed. At this moment, when I was sick, it was a difficult one. More than ever before, I was conscious how stiff and senseless the ritual was. I closed my prayerbook. For the first time I did not say my daily prayers. It was the beginning. During the next few months, step by step, I committed all possible sins against the ritual. I did not pray any more; I wrote on Saturday; and, greatest of all sins, on Yom Kippur I ate nonkosher sausage in a Gentile restaurant. I concealed my attitude at home and kept up appearances whenever they could be observed by my father's watchful eyes. But a catastrophe had to come; all this could not go on forever.

It happened during the Christmas vacation, when I did not have to go to school. My father got up very early, as usual, to go to the synagogue for morning prayers. But he did not feel well and decided to say his prayers at home instead. He planned to use my praying cubes, since his were in the synagogue. I can now understand his feelings. He went to the bag in which they lay; the dust all over the bag was an unmistakable sign that the cubes had not been used for months. The cover of dust on such a holy thing! His son not praying any more! For many years he had labored to give his son the Jewish tradition. Jews ought to stick to tradition. And now his son, blood of his blood, had broken the chain and deceived him. Red, angry, not knowing what he was doing, he rushed to my bed and began to beat me.

From a deep sleep I was awakened by the beating. I felt a shower of sharp, painful blows. It took me a long time to realize that the pain was caused by my father's hand, falling quickly and irregularly upon my body. The transition from sleep to consciousness was slow. It took me a long time to understand that this was real, that the beating I was taking was as concrete and real as a beating can be. Still only half awake, I guessed the whole story from Father's incoherent words. He had discovered my deceit.

When I felt more awake I had but one reaction: hate, hate, hate. I would have struck back if it had not been drilled into me from early childhood that the hand that is raised against a father dries up and withers. The room became full of electric potential differences, directed against me. There was my grandfather, silent, grim and hateful as always. But for the first time in years I detected a smirk of satisfaction on his face. My mother stood helplessly by. The only advice she gave was:

"Apologize to your father and promise him it will never happen again."

My older sister threw oil on the fire, claiming that she had always suspected my deceit. My younger sister had a worried and unhappy face.

Hate, hate, hate against the whole room, against my whole environment. I wanted to run away and leave everything behind me. I felt superior to all of them, suppressing my hate and desire to be rude into a feeling of superiority.

"No!" I thought, "they are not worth answering. My only reaction will be to ignore them, to run away without a word and never return to this house where everyone wants to boss me."

My father, not getting any response, felt that he had gone too far. Grim and determined, he waited for my reaction, unwilling to make a peaceful gesture. My mother repeated her pleas for an apology. Still I did not say a word.

"They don't understand me and never did. I was always a stranger here. Enough of pretending."

With this thought I dressed quickly, to prepare my only retaliation. Nobody knew what I was going to do. Silently I completed my dressing and rushed out, banging the door loudly behind me.

On the steps I noticed that I had committed two great blunders. I had forgotten to take the little money I had left on the table. The second blunder was even more grave: I had forgotten my winter overcoat. Too proud to return, I decided to go without money or overcoat. My death from cold and hunger would be a fitting punishment for them. I went out without breakfast. It was freezing outside. A cold wind blew through the streets of Kazimierz and I began to shiver. Confused by hunger and cold, by the sting of the beating which I still felt, by humiliation and hate, I could not think clearly or decide what to do. The picture of a warm room and the taste of hot coffee and fresh, crisp rolls danced before my eyes and in my mouth. I heard a voice behind me:

"You must be crazy to walk without an overcoat on a day like this. You will catch cold!"

It was a classmate from the public school, a fat, good-natured boy. I was ashamed to give my reasons and quickly concocted a lie:

"I've made up my mind to harden myself and to walk for half an hour every day without a coat. I hear it's very healthy to do so. By the way, there is something I have to buy and I forgot to bring any money. Could you lend me ten cents?"

He put his hand in his pocket and took out some change. Finding a ten-cent piece, he handed it to me. I took it quickly.

"Thank you, I will give it back soon. Now I must run, it is cold."

At the nearest shop I bought two rolls, went in the hallway of one of the houses and ate my breakfast. I still had eight cents in my pocket. Although my hunger had diminished, the cold was still unbearable. My hands and cheeks turned from white to red and from red to blue. I had to decide where to go. I had only one friend to whom I could tell the story. He was a few years older than I and lived by himself in a room near by. I was sure he would sympathize with me if I told him my tragic story. I went to his room and learned from the landlady that he was out of town and would return at six. It was then nine in the morning. Nine long hours. What was I to do with them? I could not walk through the streets any longer. The cold and wind were unbearable. My ears were numb. Brooding over the problem, I rubbed my ears with cold hands. The nasty feeling that my parents were unhappy and wondering where I was gave me strength. I still remembered vividly my father's blows.

Where to go next? Suddenly an idea struck me. Near where I stood was a Jewish library, quite a big one, clean and warm, and it opened at nine o'clock. It was only five minutes' walk away. I was saved. Holding my hands in my pockets, I ran past the staring people on the street. Having reached the library, I took a fat book and pretended to read. But disordered thoughts ran through my mind.

"I won't freeze. It is warm here. I won't starve. I have eight cents, which means eight one-cent rolls. I can easily do without anything else until six o'clock. Then I shall open my heart to my friend."

How I needed to talk to someone! Still eight hours! My anger and hate began to melt in the warmth radiated by the stove. It started with my mother. Yes, my father ought to be punished. But mainly I was punishing my mother and not my father, who was the stronger.

"My mother is worried; she does not know where I am. She only knows that I ran away without an overcoat. Now she is thinking about me and crying."

But then again my thoughts turned in an opposite direction:

"What does my mother think when she cries? She thinks that God has punished her with a bad son. But why am I bad? Because I did not pray. Why must I pray if there is no sense in it? Must I be untrue to myself to please my father? Why can't I leave home and live as my friend does? We could have a room together. I already have a little money in the bank, which I earned by tutoring. I could earn more. It will be enough for the first two months. But the bankbook is in my father's safe and he will not give it to me. This is hopeless. I must wait until I talk with my friend. I hope he will tell me: 'Stay with me for the next few weeks and we shall manage somehow.' "

Finally the nine long hours of waiting were over, ten rolls eaten and the ten cents spent. Shortly before six o'clock in the afternoon I went to my friend's. It was still very cold but the freezing wind was milder. I waited for half an hour more in his warm room, looking through the closed window. It seemed that he would never come and bring my loneliness to an end. Suddenly I saw him turn from a side street and approach the room. When he opened the door I began to cry under the strain of the difficult day.

"Pull yourself together. What happened to you?"

The only thing which he could gather from my incoherent words was that I had had difficulties at home and that I had run away.

"Relax a little; I shall prepare some tea for both of us and you can tell me your story."

He put two glasses of tea on the table, bread, butter, cheese: luxurious food for which I had been longing all day. Mingling my tears with the hot tea, I told him the story of my long day.

He adopted a fatherly air and said:

"It is a pity that I was not here. I should never have allowed you to do it. I understand that you are upset, but you must also understand your father. My father would have acted in the same way, and I should never have dreamed of running away because of that. If you had apologized and said a few nice words, everything would be finished. The best and the only thing is to go back now, to say that you are sorry, and settle the whole business. Why can't you understand it from your father's point of view?"

I refused obstinately:

"Never in my life will I apologize to my parents for not being a hypocrite. I don't want to go back. No, I don't. If you won't allow me to spend the night in your room, I shall spend it on the street. But I won't go home."

My friend tried to calm me and assured me that I could stay in his room for the night. It was my first night away from home. Exhausted by the difficult day, I slept well. In the morning I felt calmer. I even found some pleasure in preparing breakfast, in taking part in an independent life. My friend insisted that I go home. I also saw that just then there was no other way out. But I stuck to my decision: I would not apologize.

"If they insist on an apology, I shall leave home again, but this time I will take my overcoat."

It was cheap heroism; I knew that nobody would insist on an apology and that my mother was worried to death. We decided to go together. My friend offered to have a talk with my parents before he delivered their missing son.

My mother opened the door. Her face was pale and there were rings beneath her eyes. She asked me to come in and did not say another word. My friend reduced his whole speech to one sentence:

"Here is Ludwik; he spent the night in my room." Very much embarrassed, he quickly left our apartment.

My mother was too upset to show me any warmth or tenderness and much too afraid to hurt me. She said nothing but tried to give me the best food to make up for the day before. My father looked very much depressed. When I met him he did not say a word, and for the next few days we kept aloof from each other. I was treated like a stranger.

My grandfather walked through the room murmuring to himself:

"If he had been my son I would have broken all his bones and never let him come back."

My older sister thought that I had done something disgraceful and for a long time afterward called me "tramp" in our frequent quarrels. Only my younger sister and our servant sympathized with me and told me how they had tried to find out where I was, to bring me the winter overcoat.

The first attempt to dispel the atmosphere came from my father. He started with trivial remarks about passing the butter or serving me another piece of meat. I sensed his willingness to restore peace but decided not to give in easily. Never again did I pretend to pray. Ostentatiously and cruelly I ignored the religious duties in the defense of which my father had beaten me. I thought with a stiff logic: "Now Father accepts my behavior and seems to be little worried. Then why all the fuss? Is he a sadist who has to beat his son without reason?"

It was a week or two after I had returned home. The atmosphere was still tense. We had our evening meal during which I sat silent, looking at the plate and nursing the thought that I was different from the others. My father took out of his pocket a small parcel and, with shy gestures which contrasted with his usual self-assurance, slowly unwrapped it. I was puzzled by my father, who tidily put aside the brown paper and the piece of string and took a glance at my mother's face which radiated satisfaction. The inside of the parcel was wrapped in white tissue paper. When this was unfolded three shining new silk ties appeared. My father said:

"I have seen that your tie is worn out and bought you a new one."

This was an electric spark which leveled the potential differences. I thanked my father and he added simply:

"May you wear it long and in a good health."

From then on there was peace.

Why did my father change his attitude toward me? Why did his tenderness and affection increase later with the years? Why did he later help me shape a future so different from the one he had chosen for his only son?

I believe I know the answers now. I am convinced that my father, had he been a scientist, would have been a better one than I am. Without having had any formal education, he sensed from his leather carving that the ratio of the circumference of a circle to its diameter is constant and tried to determine this ratio by measurements. Without knowing the theory, he often solved correctly our mathematical school problems by clever tricks. When once during our bath together I explained Archimedes' law to him and why our legs seemed shorter in water he understood everything perfectly and thought out new applications of the principles he had just learned. Sometimes he remarked: "It is a pity that I did not learn all this when I was a boy. When I was twelve my father died and I had to work for my mother."

Through his whole life he had suppressed his hopeless rebellion. Having accepted his own hard fate, he tried to convince himself that this was the right kind of life. He needed this conviction to feel satisfied and happy. My rebellion awakened in him fears and thoughts of his own lost opportunities. Anger and cruelty directed against me were his outlet. Then his attitude changed. He saw in my fight the reflection of his unfulfilled desires. He sensed that I might succeed where he had not. Even later his attitude moved up and down on the wave line of these two conflicting emotions. But only rarely were anger and disappointment stronger than pride and friendship.

We are taught: "Listen to your father and mother because they want to make you happy." With more justification I could say: "Rebel against your father and mother and you will make them happy."

IV

IN ONE of the popular books I devoured hungrily I read about Galileo, who entered a church and saw a swinging chandelier. He was then seventeen years old. Galileo compared the rhythm of his pulse with the rhythm of the swinging chandelier and found that the period of the oscillations remained constant though the amplitude became smaller and smaller. Thus a new law of nature was born: the independence of the period of oscillations from the amplitude. If Galileo saw in the laws of nature the beauty which reflects God's glory, his ecstasy must have been intensified a thousand times more by this discovery than by the deepest religious contemplations.

In the same book I saw a picture of the leaning tower of Pisa. I followed Galileo as he climbed up the tower, carrying heavy objects which he later threw down, to become the first man on this earth to discover that they all, regardless of their weight, fall down in the same time when dropped from the same height. This law, so simple that it is nearly obvious, is only three hun-

dred years old! I learned that this and other laws belong to
physics, a science which deals with the phenomena governing
inanimate nature.

Once a cousin visited us from a provincial town. His mother
sent him to see the great city of Cracow as a reward for passing
the matura. I asked him how he liked physics. He said:

"Physics and mathematics were the dullest subjects. I forgot
everything that I learned an hour after the matura. Even biology
was much better. I was never interested in physics and mathematics. But there was one half-crazy chap in our class who was
a genius in these subjects. He bought three big volumes of
physics and knew everything better than the teacher."

I was astonished. Is there enough physics to put into three big
volumes? If there is, I could get the volumes and learn absolutely
everything about physics. I asked my cousin whether he remembered the name of the author. Yes, he remembered. They
were written by A. Witkowski, a professor at the University of
Cracow.

The next day I searched in the secondhand bookshops. I found
the three heavy volumes, seven hundred pages each, and after
long bargaining they became my property. I did not know at
that time that this excellent book in which the principles of
physics were explained with masterly simplicity and clarity was
a basic textbook used by the university students in physics.

It was hard reading. No stories, no pictures, no technical
details; only the principal laws and basic experiments. After
some twenty pages I was stuck. The symbols *cos*, *sin* appeared.
I did not know what they meant. I saw that to understand
physics one had to know more mathematics than I did. Again
I went to the old narrow street in Cracow full of secondhand
bookshops, in search of a mathematical textbook.

With pain and joy I made my way slowly through the pages
of the book. These pages helped me to build an impregnable wall
around myself. I saw in science an escape from reality, a source
of emotion in the glow of which even my grandfather seemed
only a harmless and unhappy fool.

The references in the book led me to new ones. I started to

collect a library in physics and mathematics and spent on books all the money I could get from my parents or earn by tutoring. Slowly I realized that science is not a sealed book but an organism pulsing with life, changing and developing. I began to sense that I had gained only the first foggy glimpse of the great land of promise. I wanted to learn physics and to become a physicist.

I was never attracted to physics by its engineering side. Never in my life have I shown any mechanical ability. My hands never worked, I am ashamed to admit. (I am always so ready to be ashamed of my idle hands that I suspect myself of being really proud that I did not stain my hands with work.) What I loved in physics was the rigorous character of its reasoning; it seemed most wonderful to me that so many complicated facts can be deduced from so very few simple principles.

I wanted to study physics and mathematics. But to study one must attend an institution of learning: the university. To attend the university one must pass the matura. I came back to the old problem of changing the fate that had been designed for me by my father when I was ten.

To pass the matura is not simple. I was in my sixteenth year. I had only a little more than two years' time if I wanted to take the matura at the proper age. According to the rules, I could enter the examination without any studies in the gymnasium, but then the examination would be much more severe. All the knowledge had to be packed carefully in my brain and reproduced in one week's written and oral examinations. Besides, I had only afternoons free; in the morning I attended the business school. But once the idea of the matura possessed me it did not leave me in peace. Every time I passed the university building I was moved by a feeling of longing and suspense. Would I be able to enter it as a student? The first thing was to collect information: what must I know for the matura? With the exception of Latin nothing really frightened me. According to the new rules, Greek could be replaced by French, of which I knew a little from the business school. Mathematics did not scare me, and I knew more physics than was required. I had read enough to know sufficient Polish literature. History could be learned from books. In this way I went through the program of the gymnasium to decide whether or not I should be able to make the grade. I decided to try. The first thing was to buy a second-hand textbook of elementary Latin. The absolute uselessness of my work was amusing, but the knowledge that each lesson brought me nearer to the university was exciting. My older sister, who had the idea that I never stuck to anything, ridiculed my first attempt to learn Latin and said spitefully:

"Now you buy Latin books! After you spent money for three volumes on physics you try to find an excuse not to read them. Oh yes, you will pass the matura! I am sure that you will pass the matura with honors!" putting strong emphasis on the last word.

My grandfather was very much amused. He did not know very well what it was about, but anyway he was certain that I would not amount to anything. So he said:

"You will get a gulden from me if you pass the matura," and murmured to himself in good humor:

"I can just imagine him doing anything in his whole life!"

I devised a careful plan for the next two years' work, leaving plenty of time for repetition, reading and deepening my knowledge by other books. When I came to the actual work I found it easier than I had thought.

After learning aimlessly in business school it was quite different to study with a well-defined purpose. The information acquired went smoothly to the proper places in my head and remained there. I was unsystematic and disorderly in my schoolwork. But here, in taking notes, solving problems or translating Latin, I was orderly and systematic. When I began to read Ovid, Virgil and Cicero my older sister ceased to make ironical remarks. On the contrary, she declared her willingness to help me in French and did it with enthusiasm and loyalty. I could not help teasing my grandfather about the gulden which he had promised me, but his only answer was:

"I don't care whether you break your neck or pass the matura, and I never promised you anything."

My father began to watch my studies with sympathy, and my mother worried lest I was overworking myself and damaging my health.

At the same time clear and decisive ideas about my future and how I was to earn my daily bread began to take shape in my mind. I wanted to be a teacher of physics and mathematics in a gymnasium in Cracow. To achieve this seemed to me the highest level of happiness.

At this time I knew quite a lot about anti-Semitism theoretically but very little practically. The ghetto formed an isolating layer around my life, and the impact of anti-Semitism did not reach me. But I knew theoretically how difficult it was for a Jew to obtain a teaching position in a gymnasium. And this is what attracted me. My desire to be a gymnasium teacher must have been connected with the experiences of my childhood. To be a Jew and a medical man or a lawyer meant to have Jewish patients or Jewish clients, meant to have an office near the ghetto and to depend on the ghetto. But to be a gymnasium teacher certainly meant an escape. I would probably be the only Jew in the gymnasium. What a wonderful feeling to be the "only Jew." A Jew is very proud to be the only one somewhere. So should I be. Nourished by the events of my childhood, my subconscious desire to escape played an essential part in shaping my future.

There was, perhaps, one more reason. I had never gone to a gymnasium and always felt unhappy about my education in a citizen's school. I might pass the matura, I might even study at the university, but the experience of the gymnasium was lost to me forever. I must have felt a yearning to enter as a teacher the school which was closed to me in my youth, to command and to instruct the boys in blue uniforms whose lives I had envied in my childhood.

V

THE PEACE-LOVING OLD EMPEROR decided to lead Austrian troops to a victorious war against Serbia, then Russia, then France, then England, then against the whole world. "God punish England" was the patriotic slogan, and the spiteful tongues added: "With Austrian management." On our side were God and the invincible German troops.

I was glad of the exciting show. War, I imagined, would be a short spectacle, finished after six weeks, not touching my own life but changing the world around me. I was too young to be an actor on the bloody stage but old enough to enjoy the performance from a comfortable seat on the side lines. We all nursed the same hope: war would create an independent Poland; the dreams of the great Polish poets would come true. And it did not even matter who won. Both sides promised the restoration of our country.

Our leather shop boomed. Every evening my father came home with a bag full of paper money. Prosperity entered our home.

I worked, according to my schedule, on Latin, history and other matura subjects. The world might blow up and be drowned in blood and hate, but I should still have my matura.

Months passed; the show began to be dull and prolonged, with no end in sight. I moved through events which were, according to newspaper reports, full of the glorious deeds of our army. The stream of time carried me to the noble duty of defending the fatherland.

My oral matura was scheduled for May 9, 1916. The same year, in April, posters announced the mobilization of my class;

all men born in 1898 would be examined. I went to the proper barracks, undressed and joined the queue of completely naked men. We moved slowly forward, our bodies touching each other. I stepped on the scale, and an orderly wet my chest with a sponge and wrote down with a blue pencil a number indicating my weight. From there I was pushed by the queue to a measuring stick, and a second number marking my height was written on my chest. Six feet tall. The attending soldier smiled with satisfaction. I was ready to appear before the commission, which glittered with silver and gold collars. The military doctor looked at me approvingly, put the stethoscope to my heart for a second, then to my lungs for another second, and I heard the word *tauglich*—fit.

I dressed, went to a room where I had to wait until all the soldiers had been examined and assembled. The important moment came. A sergeant entered the room and commanded:

"Raise two fingers of your right hand and repeat what I say word for word."

In a grim chorus we swore obedience, loyalty, readiness to sacrifice the last drop of our blood for the Austrian emperor to defend our fatherland on land, sea, air, and to fulfill all the commands of all our superiors. From then on we were soldiers. In the papers which I obtained it was stated where and when my service would commence. The date was May 11.

On the ninth of May I took the oral matura. The first question was in history: "Poland's fights for independence." The picture which I drew had two distinct colors. Black, completely black, for the Russians and white, completely white, for Poland. The chairman looked with interest, the professor with satisfaction, when I poured out smoothly details of Poland's part in the Napoleonic Wars, the dramatic story of the two Polish wars for independence in 1831 and 1863, and finally, raising my voice and sounding full of emotion, I concluded:

"The struggle for Polish independence is not over. Pilsudski's legion, arm in arm with the Austrian army, fights the brutal force which destroyed our independence but could not break our spirit. Each day brings new signs of Polish heroism, each

day increases our hopes and each day we pray to God that great
and powerful Poland will again be free, that freedom and happiness shall grow on this earth, saturated with the blood of
Poland's best sons."

This sort of thing was loved at that time. Words are cheap.
I myself was deeply touched by my own speech.

From the fight for Polish independence I jumped straight into
the equation of the ellipse and its properties. The teacher in
mathematics was intelligent and soon saw that I understood and
knew the material well. Everything else went very smoothly.
The result was that I passed the matura with first honors.

One day was left free for me to recover from the emotional
strain. The matura was supposed to have marked the end of an
old life and the beginning of a university career. Instead it
marked the end of my civilian life. A day later I went to the
military barracks, and among the few things which I took with
me was a book on differential calculus.

The barracks were a few miles outside the city limits of
Cracow. I entered the barracks and saw a group of civilians
looking idly at rows of drilling soldiers. I joined the civilians,
trying to make conversation. In an hour or so, I learned, we
should get uniforms. The sergeant directing the drill noticed
us and with a disgusted face shouted:

"You civilian riffraff, you damn future recruits, stand away
and don't take so much space. The courtyard is for soldiers and
not for lousy civilians with milk under their noses and filth in
their trousers."

Obediently we formed a line, pressing our backs against the
dirty walls of the buildings which framed the big court. The
commands, "Quick march, halt, quick march, halt, right turn,
left turn," rang out in a loud staccato. The German words of
command were mixed with juicy Polish and Ukrainian oaths.
It reminded me of the pedagogical methods used in the Jewish
school. The picture of an old bearded teacher replacing the
sergeant came—for the first time—as a memory of a pleasant
dream from a world which had passed out of my life.

In a near-by rank a soldier went one step farther than he

should have before halting and turned left at the command to turn right.

The sergeant called him:

"Hyrko Wasilewicz, come here. You damn bastard. You don't behave like a soldier, you act like an old whore on pension. This is a military barracks. How many times have I to tell you where your right hand is? When you eat your food, you greedy dog, you know what hand to use, don't you? With what hand do you take your food, you son of a whore?"

The soldier stretched out his left hand. The pedagogical methods acquired by the sergeant did not cover left-handed morons. He began again:

"It is a pity that your father was not castrated twenty years ago. He and your whore of a mother wouldn't have produced such a bloody bastard as you are. Say: 'I report obediently I am an idiot.'"

The soldier raised his hand to his cap in a salute and repeated the sentence with an indifferent expression.

I lost the feeling of reality. It took me months before I understood the peculiar language of the army, the fact that words spoken there had lost their literal meaning, that their edge was worn out and softened by constant use. When I learned this I also learned to speak the same language. Its use made life somewhat easier.

Soon the uniform depot was opened. We got summer uniforms which had gone from one generation of soldiers to another since the start of the war. They were indefinite in color and shape. They had only a definite smell: a mixture of dirt and strong disinfectant. The first man, shocked by the bundle of stinking rags, asked for something better. The corporal leered.

"So that's it. I see. So you don't like the bloody uniforms. When you go to the front you get a brand-new fine uniform. And if you have good luck you will be buried in it while it is still good-looking."

I took my bundle without protest, went with the others to a room with iron bunks. This room was to be my home and that of thirty-nine other soldiers. In the middle of the room was an iron stove and a table with two benches. These and the beds comprised the furniture.

The regular military drill started next day at five o'clock in the morning.

"Right turn, left turn, left, right, left, right, attention, number, quick march, halt, on the hands down, on the feet up, down, up, down, up, down, up, dismiss, fall in, smarter, smarter, smarter, smarter, only one sound, all at once, down, up, down, up, this is an army not a brothel, smarter, smarter, I will teach you a lesson for the rest of your lives, I will keep you in prison until you rot, smarter, you sons of bitches."

Each soldier who had passed the matura wore a yellow stripe on each arm. They were called –officially and in all seriousness– "the sign of intelligence." All forty soldiers with the yellow stripes were squeezed into one room in which some of us played cards in the evening by the light of a candle in a beer bottle, drawn together by our mutual hate of our superiors, quarreling wildly among ourselves, using the language which we absorbed automatically by breathing the heavy air of the military barracks.

Once when repeating the "ups" and "downs" I saw the company commander approaching our platoon. His eye moved slowly from one soldier to another, then rested upon me for a conspicuously long time, with the result that I became even more confused and behaved like the wretched Hyrko Wasilewicz whom I had seen the first day. The captain pointed his finger at me.

"You soldier, come here."

I came nearer, saluting the officer.

"What's your name?"

Saluting again, I said, using the soldier's staccato as well as I could:

"I report obediently, my name is Infeld."

He put his arms akimbo and looked at my yellow stripes.

"Ha! So you have been in a gymnasium. And you are supposed to be intelligent. You are the worst soldier in this platoon.

You are a disgrace to your fellow soldiers. And what an example you give to the whole company! And you would like to be an officer in the future! What kind of an officer will you make if you cannot do better than any peasant idiot? This is not a university, this is an army where you must think quickly and smartly. Dismiss."

At least he did not use the sergeant's language. This was some relief.

Day after day I repeated the same drill but made no real progress. I was tall, well built, but clumsy. My muscles were scarcely developed. It was not my fault. Never as a boy had I taken any part in sports. I had never had any playtime in my youth. I was astonished at how easily my fellows soldiers learned to repeat the brisk motions and how difficult it was for me.

But it was not only physical clumsiness. The business was so boring that I found escape in daydreams while performing the mechanical drill. Often, not having heard the order to turn about, I marched ahead a few steps, suddenly finding myself alone, to the amusement of the other soldiers and the livid anger of my superiors. There was a word in the Austrian army coined for such soldiers as I. They were called "oferma." With humiliation and helpless anger I found myself the first-class "oferma" of our company.

One of the first things which we were taught was how to salute our superiors. All superiors, whether in the barracks or on the street, had to be saluted. Whenever I saw stars on a collar I had to turn my heard sharply, looking straight into the eyes of my superior. Then I had to raise my right hand very smartly to my soldier's cap, still looking into the eyes of my superior. In this position I had to pass my superior, starting the performance three steps before I cut his field of vision and keeping it up for three steps after I left it. Then, though theoretically I was not seen by the officer, I had to lower my hand very smartly. In practice this meant that a walk was for the ordinary soldier a continuous, deadly boring performance as he went through the streets full of officers.

About this time I went through the ordeal of my first romantic love for a girl whom I had coached in mathematics, and I spent all my free time with her. Impelled by a strong desire to overcome my shyness and fear, I went at an amazingly slow tempo through the process of kisses, touching her face, then neck, then breasts, fighting my own feeling of guilt rather than her feeble resistance.

The day after seeing her I always dreamed, during the mechanical drill, of kisses, of the fresh smell of her body, of her tiny but well-formed breasts, counting the hours dividing me from our next meeting. Sex was for me a mystery wrapped in a veil of silence, experienced only in its preliminary acts and most intimate thoughts.

During my free hours we sat together in a dark corner of one of Cracow's parks, my arm feeling the warmth of her body through her thin summer dress. Then suddenly I would have to rise smartly and, in a standing position, smartly salute a passing superior.

Sometimes a pass was refused. Then I had to spend the evenings in the barracks listening to the talk of my fellow soldiers. It spun around one subject: sex. Most of my comrades were from outside Cracow. When they were allowed to go out there was only one place to which they went together: a brothel. For hours afterward, I had to listen to a detailed account of their experiences.

My life had changed so suddenly and completely that I lost the feeling of my own identity. This feeling, based on the connection between past and present, collapsed under pressure of the contrast between my life of yesterday and today. I did not realize that it was still I who was pushed through the world of changing events. In the evenings, when my fellow soldiers played cards and discussed sex, I opened my mathematics book, seeking a bridge to my past. But the physical strain of the day and the depressing atmosphere were too much for me. My brain refused to respond.

There is one peculiar fact which I can hardly explain. I was not disliked by my fellow soldiers. They rather pitied me. Dirty tricks were played on some "mama's boys" but not on me. One evening the leader of our group, the toughest fellow among us, came near, looked at the book I was trying to read and said:

"My God! This man must be a genius or a crazy bugger." Afterward he was very kind to me and even tried to give me hints on handling my rifle and avoiding the wrath of superiors.

A few days after my new life had started at the barracks my father visited me. We were not yet allowed to go to town because we had not learned to salute smartly enough. Suddenly confronted by the visible connection between my previous life and the misery of that day, I started to cry. I did not even try to keep back the stream of tears; with their salty taste in my mouth I repeated:

"I cannot bear it. I try my best, but it is unbearable. The front must be better than this hell."

My father was unhappy but more cordial than ever before. I sensed that he was afraid I might commit suicide. He said with embarrassment:

"You are my only son. Mother is not very well. She could not stand it if anything happened to you. You will see that God will help you. You were always a good son." Then he took a new watch from his pocket and added: "I did not buy you anything for your matura. Here it is. It is a very expensive watch, the only Swiss watch they still had in the store."

Before going away he told me, "You can always count on your father. I will do everything to help you."

What he meant was that he would try to get me out by bribery. Corruption in the Austrian army was a trivial matter, and nobody who had money hesitated to use it. But in exceptional cases bribery was not possible. The doctor in our company was such an exception. Spiteful tongues claimed that he had taken money before, nearly faced a court and was scared.

In the fourth week of our drill we learned during morning parade that in one of Cracow's café houses a soldier had not behaved properly and had been impudent to his superior. The commandant of the military district of Cracow decided to teach all soldiers a lesson. They were not allowed to leave the barracks for ten days. Our captain, after reading the order before the assembled company, made a speech which concluded:

"And to the ten days of barracks arrest I add five days for my company, so that you will learn for the rest of your lives that a soldier must be a soldier—and so that the whores in the brothel may have a holiday too."

I wrote to my girl breaking the appointment we had made and telling her the reason. She had the courage to come all the way to the barracks, and the sergeant allowed me to go outside to see her.

She brought me a book to read, told me how much she loved me and how everybody has to suffer. She enjoyed expressing herself in high-sounding phrases which at that time I swallowed with delight.

"You know, don't you, that you are my darling and that I love you with all the power of my burning heart. We are all drowned now in a sea of misery. My poor mother cries her eyes out for my two brothers who are at the front. Fate has struck us all hard."

Touching her hands, I said:

"Why does no one revolt? You cannot imagine how we are treated here. Worse than beasts in a circus. We are tortured physically and mentally. I can't stand it any more. And the world could be beautiful if it were governed by love instead of hate."

"Try in your moments of despair, my dearest, to remember and think of somebody outside these cruel barracks. Somebody who thinks constantly about you and who understands you, who is full of desire to help you and make you happy."

We were carried by our bombastic phrases into the thin air high above the barracks where the sight and smell of those gray sad buildings could not reach us. But my father's Swiss watch brought me back to earth. It was time to return. We kissed each other passionately and I went to the dirty room, undressed, went to bed, trying not to hear, not to talk, but to preserve by silence the freshness of my emotions and memories.

A corporal came to check whether we were all lying in bed. He turned toward me:

"I see that you have quite a nice girl. Why didn't you take her into the bushes?" Then he described how, in my place, he would have done this and that, so many times and in so many ways.

I heard my heart beating so loudly that I thought it would break. Through my tightly squeezed lids I visualized the obscene gestures that accompanied his words. More than the corporal, I hated myself for not having the courage to slap his dirty face. The half-drunken corporal went out. I suddenly found relief in a spasmodic cry. Everyone, even the toughest men, tried to show me their sympathy. One of them said:

"What do you care about that son of a whore? Wait until we are lieutenants and we shall show these bastards how to behave toward soldiers with 'signs of intelligence.' In civil life he looked after pigs and here he wants to show off."

The next day, under all this accumulated strain, I fainted during drill. I felt that I could not go on, come what might. I lay on the grass and did not move when the captain came near.

"Tomorrow you go to the doctor, and beware if you are a malingerer. I will put you in prison until you rot."

Next morning I registered for the doctor's visit and went to the small room where the doctor received sick soldiers. I was the only soldier with yellow stripes. The medical officer and his orderly sergeant came in. The doctor was small, fat, with a red mustache and an expressionless face. He hung his sword and black cap on a hook while the orderly placed us in a line according to the time we had registered. I was third in the row. Then the orderly called the first soldier and sat down to write the doctor's verdict.

"Step up."

It was a Jewish soldier. In Polish-German-Yiddish he described how his head ached and everything danced around him.

"One aspirin tablet. Fit for duty."

"Next."

A soldier showed his red swollen feet.

"One day off duty. New shoes of larger size."

"Next."

My turn came. I heard the beating of my heart. I mumbled my complaint incoherently, partly through excitement, partly by design. The doctor listened to my heart. I thought:

"My future, perhaps my whole life, depends on the few words which the doctor will say now." This thought raised the activity of my heart and I heard its violent response. By pure thought I tried to excite still more my arhythmically beating heart. I pictured myself beaten, humiliated by the corporal and sergeant and killed in action. My heart responded still more strongly, intensifying its beat and confusing its rhythm.

The doctor turned toward the sergeant.

"What a peculiar rhythm, listen," and he handed the stethoscope to the orderly, singing out: "One, two—three; one, two—three."

The orderly smiled and shook his head approvingly. The doctor dictated:

"To the garrison hospital for a heart examination."

In one second the sun began to shine over my head and the world was again a glorious place. There was hope. The scene would change. I should go to the hospital where there was no drill. Away from the barracks! I was convinced that my heart had not been able to bear the strain of the last weeks and that I had some kind of heart disease. But a heart disease at that time was a great treasure of which everyone who had it was proud. My only worry was that I might get better before the war was over. Even if one dies of a heart disease, it is still different. It is a calm, comfortable death in bed, with doctor, nurse, everyone crying, an honest civilian death. I loved my sick heart. I could have taken it out to kiss it for joy and to beg it to remain bad.

The organization of the Austrian army will always remain a symbol of the height of stupidity and red tape. In this army the process of examining my heart and deciding what to do with me took six weeks. The final diagnosis was:

"Inflammation of the heart muscle. Capable of auxiliary service but not fit for active service at the front."

I was sent to the fifth company of our battalion, a company

for soldiers not capable of active service: cripples, malingerers, old soldiers and idiots.

These dirty wooden barracks, full of lice, rats, bugs, seemed a comfortable heaven compared with my previous environment. The commanding officer attached me to the company orderly room as a clerk, remarking contemptuously:

"I guess that if you have the matura you can write and add."

Nothing could disturb my happiness in the first few weeks. There was no cruelty in the company. The commandant was an old officer taken from the reserve, barely competent to watch a pack of cripples and idiots. Soon I was allowed to sleep at home and come to the office early in the morning. Nobody cared if I left the office during the day for a few hours. Most of the time I sat with a book before me, waiting until someone gave me a job of rewriting, stamping some paper or drawing lines. Whenever I liked I could walk through the barracks, visit the canteen or the soldiers' sleeping quarters with their peculiar smell: a mixture of dirt, sweat and the unforgettable odor of squashed vermin.

Once, when passing through the court, I saw a young soldier leaning against the wall and reading a book. It was an unusual sight. I thought that no one in the Austrian army beside me was interested in books. But my astonishment increased when I came nearer and read the title. It was the Polish translation of *Science and Hypothesis*, a brilliant popular book written by the great mathematician and philosopher, Henri Poincaré.

I tried to appear casual though I burned to start a conversation. I said:

"That is an excellent book."

My neighbor's look expressed astonishment.

"Do you know it?"

"Yes, I read it."

In a deep voice he talked nervously about the book as though this were the only world problem which really mattered.

"I cannot believe that the laws of nature are only conventions." He paused for a moment, trying to clarify his doubts. Then he added, moving his hands nervously:

"Take something, say a stone. And then you drop it. Then, of course, the stone falls. It is called accelerated motion, if I remember. I don't see how there is any place for a convention. I don't know whether you understand what I mean."

We started to walk around the square court. I tried to defend Poincaré's view.

"It is a convention, at least to a certain degree. You are right that the experiment tells you that a stone falls down with a uniformly accelerated motion, but only when you use a good clock. But imagine that you use a crazy clock. The law is all right if the clock is all right. But you can imagine a clock which changes its rhythm irregularly. Then, for an observer who uses such a clock, the motion of the falling stone will not be uniformly accelerated."

Our conversation proceeded in this strange setting of a military barracks. My neighbor, whose name I did not know, argued:

"I don't see. There is some psychological feeling of time. Some kind of intuitive feeling which we all have. This feeling would tell you that a wrong clock is wrong."

I tried to draw the discussion back to the safe ground of physics.

"But if you talk about psychological time you thrust the argument into philosophy. I don't know what psychological time means. There is a beautiful example given, as far as I remember, by Poincaré. Imagine that we woke up one day and all the clocks in the universe had slowed down their rhythm, say by half. Imagine that everywhere the rhythm was slowed down, even the rhythm of our hearts. Then we could not detect any difference. Therefore the whole example is meaningless."

"No. This cannot be right. Imagine somebody awakening suddenly. He awakens in the night. It is dark. His sensations would change suddenly. Then he would have also a memory of the past. We always carry our past with us. No one can get rid of his past. What I mean is that such a man would have a feeling that something is changing in the world."

He spoke tensely, in jerky sentences, pausing suddenly in the middle as though it were too painful to finish a sentence. I asked:

"Would not such a man, awakening suddenly in the night and feeling some heart trouble, rather believe that he had a bad dream than that the clocks in the whole universe had changed?"

My companion did not answer for a while. Then, changing the subject, he said:

"There are many things which I don't understand in this book. I don't know what a differential equation means. If I am at all interested in the whole thing . . . assuming that I am interested . . . which is doubtful . . . then only in the philosophical part. But even this matters little . . . really."

Here in these barracks the most unexpected thing had happened. I had found someone with whom I could talk freely. I was anxious to make our discussion more personal.

Soon I knew more. His name was—ironically enough—Samson. He was thin, small, weak, sent to the fifth company because of lung trouble. Samson was a year older than I. He told me hesitantly and chaotically about his family. His father was a son of a rabbi in a small Jewish town in East Galicia. At fourteen he had married a girl chosen by his parents. He saw her for the first time at the marriage ceremony. Samson was born when his father was sixteen. Life was designed for him by a long chain of tradition. Samson's father had studied to become a rabbi and Samson would be the next link in this chain; the oldest son must follow in the footsteps of the father.

By accident some books other than those of Jewish learning came to Samson's father when he was eighteen. They changed his life. Some years later he passed the matura, and when I met his son at nineteen he had just started a law practice in Vienna. I thought how insignificant and trivial were my difficulties when compared with those of Samson's father.

We spent all our free time together and our talks grew more intimate. Once we went to a village saloon near the barracks. Samson ate and drank his beer quickly, behaving as though it were an odious job which must be done. Between bites and sips he complained:

"It is idiotic. . . . Why do I go on living? Death is peace, the end of everything. . . . I wish I were dead. Why don't I commit suicide? Because I cannot bear the cry of my mother. Mother love is primitive, fundamental. What does one want from life? A clean lavatory with flushing water and not the filthy latrines we have here."

"You want to shock me. You don't believe in what you are saying. There must be some sense in life. If nothing else, then there is science and human kindness. Even here, in the most depressing and foul atmosphere, you find some human kindness. Did you meet the small Jewish corporal in the office? He is one of the kindest men I ever met."

His face relaxed for a moment.

"You are interested in outside life. I am not. I am not interested in anything. I may read books or write poetry, but it is all dilettante. I don't feel reality around me . . . emptiness. Sometimes perhaps I feel it. But then I hate it."

"Don't you have longings, desires?" I asked. "Wouldn't you care at all if you were allowed to go home to Vienna to do whatever you like, to be free, to read books, to go to the theater and to walk through the streets without saluting?"

"All this matters very little. I am not even sure that I don't prefer to stay here. Everything here is foul. Everything here is disintegrating. And I am too. You would not talk to me if you knew what I did. I did things which the so-called decent man would never do."

"What did you do?"

He hesitated for a while. "I stole bread from a soldier. You know what bread means to a soldier. The only thing they eat. And they cannot buy it. I opened the knapsack of a soldier, took out his portion and ate it."

"You didn't do it because you were hungry. You wanted to experiment on yourself and show how low you can sink. You would love to prove to yourself that there is nothing so dirty that you couldn't do it."

"What you say sounds clever. But it is not so. . . . I was simply hungry. The soldier was hungry too. But I did not have moral scruples."

"Let's get out of this stuffy room." I paid for our lunch and we went outside. (I found it quite natural that Samson never bothered to pay, behaving as though he did not even notice that I did it for him.) We sat on the lawn; Samson took pieces of grass, chewing them nervously, dissecting them in little bits and talking with great effort.

"When I joined the army I thought: I shall be a good soldier. I hate the war and all this filth as much as you. But it is all a symbol!" He shouted out the word symbol. "Whatever you do is symbolic. I cannot be a good soldier. Why? Because I cannot be anything. I cannot make friends and I cannot talk. You are the first person I have talked to. But even this . . . I am sure . . . you will see, will finish badly."

"What you say is all nonsense." I tried to scare Samson by my violent reaction. "Nobody can be expected to do a good job of a thing he hates. Who are the good soldiers covered with medals? Who are the soldiers who get more and more stars on their collars? Look at them. Teachers in elementary schools in forgotten small villages. They suddenly got power over grown-up men. They are drunk with power. They are excited over the great adventures of their lives. Then you have butchers or tough boys, for whom the whorehouse is a social club. One can do things well only if one likes them, if one believes in them. Even the officers in the army know it. You know what the proper answer is if you are asked what do you clean your rifle with? The proper answer is to say: with love and eagerness. You cannot even clean a rifle properly without love and eagerness. I am sure that there must be something which you would like to do. And if you do it you will do it well."

"No. Absolutely nothing. I am sure. I read books. But it is really only intellectual masturbation. It does not mean much. I could live without them. I never get excited about books the way you do."

"Don't you feel," I asked again, "a social duty to do something? I don't want to be a businessman because I don't believe in the social value of this work and because I hate it. One must choose a work in whose social value one believes. What about the cultural heritage which we obtained? And don't forget that the flushing water in the lavatory is also included in this heritage. We have to pay for all that by our own work. We must contribute something from us. It is a debt which must be paid. And it is dishonest to ignore a debt."

Samson stared at me, then suddenly leaped to his feet:

"It's all childish, cheap phraseology. I didn't want to come into this world. Why must I be thankful and for what? To hell with your cultural heritage! Where did it lead us? Here." He pointed toward the barracks. "If I had to do something in life, then it would be manual work. I would lay bricks or cut stones or make shoes. I would not need to hear highbrow phrases. I detest them. Do you know what I find the worst thing in the barracks? Do you know what I detest most? The stinking yellow stripes. The 'intelligentsia' sign. I prefer any time an ordinary soldier, a peasant, to a guy with the matura."

We both grew angry and tried to hurt each other. I replied:

"When I was four years younger and read Rousseau's rubbish I thought as you do. Now I know for sure that it is utter nonsense. You like the simple peasant. Look what becomes of him if he is a corporal: a cruel, grasping animal. Don't you see," I shouted, "that through intellectual development only you understand other people and you see your own limitations?"

Samson became suddenly apathetic. He only murmured, "It is all nonsense," and refused to go on with the argument.

As our friendship grew Samson leaned on me more and more. I thought that it was my duty to inject some life force into his disintegrating will power. But all my attempts had, if any, only a temporary effect. I found that the best method was to cut down our talks and read books together. Among others we read Whitehead's *Introduction to Mathematics* and Russell's *Introduction to Philosophy*. We both took refuge in a world constructed by thought.

Gradually I forgot the difficult days in the old military barracks. I no longer looked at my present life through the prism of my previous unhappiness. The idle days and their boredom increased my restlessness. The beginning of the academic year came nearer. I could not bear the thought that the lectures at the university might start with me sitting in this silly office instead of the lecture room.

Once the sergeant gave me the easy task of writing a report for the battalion headquarters. I wrote it in large clear letters, so that even an officer in the Austrian army would not have much trouble reading it. The report came back with an angry remark that according to order No. 586F5/435x18W, paper must be saved and reports written briefly in small letters on paper one side of which had been used before. Therefore the report was not accepted and must be rewritten. This time the sergeant was really angry. His normally pink cheeks grew pinker and, nervously stroking his small brown beard, he shouted:

"What do I get from you? Nothing but trouble. Once in two months I give you something to do and you do not turn out a report but . . . Why don't you stay home and waste your time there? Who needs you here? Go away and leave me in peace."

I understood the implications. It was a clear hint. My father arranged the details with the sergeant. A gentlemen's agreement was made according to which the sergeant would get a sum of money from my father for each week I stayed home.

Again I lived at home. I tried in vain to suppress the thought of the risks I had taken. An unexpected inspection, an accident, might reveal that I had bribed the sergeant and deserted the barracks. My imagination raced, picturing scenes in which I was hauled before a military court-martial and sentenced to imprisonment. Then I tried to convince myself that the danger was small, that I was protected by the incredible disorder in the Austrian army.

But there was one practical problem which had to be solved: How to walk through the streets of Cracow. It was impossible to go out without being molested. I could wear my civilian clothes, but sooner or later a private detective would certainly approach me. His job was to hunt for the scores of deserters living in Cracow. Detectives stopped men of military age and, showing their badges, asked for military papers. Nobody dared to walk about without some kind of identification. Even my father always carried his birth certificate. So civilian clothes were out.

In my soldier's uniform I was free from the civilian hunting dogs; the law did not allow a civilian to stop a soldier. But in uniform I was exposed to the danger of Austrian gendarmes looking for soldiers and asking them for their passes.

Samson formed the link between me and the barracks. When I saw him my first question was:

"Is everything all right? Nothing happened?"

He was annoyed that this was my greatest concern and impatiently answered:

"No. Nothing happened."

I asked him what to do to gain freedom of movement.

His suggestion was ingenious:

"I read Chesterton's short story in which a man is dressed so that he appears a gentleman to the waiters and a waiter to the gentlemen. He tries to steal some silver."

"What happened to him?"

"He was caught by Father Brown."

"What has that to do with me?"

"You could wear something between a soldier's uniform and a civilian suit. The detectives will think you are a soldier and the gendarmes will think you are a civilian."

"And Father Brown?"

"He is in England. There is nobody so clever in Austria."

I took Samson's idea seriously. The soldiers' uniforms were so varied and untidy that one could imagine a combination to fit his plan. But there was one insurmountable difficulty: the cap. There was no continuous transition from the cap of a soldier to the hat of a civilian. They were entirely different. The plan had to be abandoned.

Next day Samson came to me and with a bored expression, handed me ten passes bearing the stamp of our company, genuine pass blanks which needed only a forgery of the contents and of the commandant's signature. I was overwhelmed, thanked Samson and asked him how he had got them.

"It was simple. I asked the Jewish corporal. He had the passes and he gave them to you."

Samson was disgusted that the few scraps of paper could make me so happy. He tried to switch our conversation to more profound problems, in which I showed little interest.

I felt like Macbeth committing one crime after another. The first time I went out with a forged pass I felt jittery. But nothing happened. Nothing happened the next time. I became more bold. I thought: "The gendarme does not know the name of our company commander anyhow. It is better to put a fictitious name than to forge a real one." So I put any name which entered my head: the names of Polish poets, university professors, friends, even the name of my grandfather, writing their imaginary ranks, "lieutenant," "captain," with the usual abbreviations. Later I was stopped several times by the gendarmes, but they always found my passes in perfect order.

The academic year began at the Jagiellonian University in Cracow. In a mood of expectation and excitement I went to the first lecture. I entered the Gothic building in one of Cracow's beautiful parks, the chief university building where most of the lectures were held. The former mathematics and physics buildings had been changed into hospitals. It was evening and the corridors were dimly lighted with gas because of war economy. The small lecture room was nearly full. The audience consisted in the main of women and a few soldiers in uniform. My dreams had come true. Only the accessories had changed. I did not know that I would enter the university in the uniform of a soldier.

The lecture was on the theory of numbers. A professor rushed in like a bombshell and talked like a machine gun, tearing his small beard, walking in great strides and being very enthusiastic. For the first time I heard of Peano's number theory as an attempt to reconstruct mathematics axiomatically, starting with definition and axioms, widening the scope of our knowledge cautiously and slowly.

The exposition of most of the lecturers was clear and interesting. At that time the chair of physics was occupied by Smoluchowski, a great and famous scientist, one of the rare types who combine knowledge and creative work in both theoretical and experimental physics. Although the subject of his lecture was known to me, some of his critical remarks were like revelations. When lecturing about dynamics he recommended Mach's book, a profound criticism of Newtonian mechanics. I bought the book of the Viennese physicist and philosopher, studied it carefully, and for the next few years I was strongly under the influence of Mach's philosophy. It was the philosophy of a scientist. It seemed simple, convincing and devoid of metaphysical obscurity. When Mach explained that scientific theories have an economic value, that they express an immense number of data through single formulae, through functional dependence, one could easily understand what he meant. I was not critical enough to see also the shortcomings of his philosophy which led him to a bitter crusade against atomic theories. But the part concerning the analysis of classical mechanics was excellent. In Mach's criticism of Newtonian physics were the seeds of ideas which later, in the hands of Einstein, developed into the great structure of general relativity theory.

The military and scientific aspects of my life were interwoven, forming a curious pattern.

One day, for reasons which I don't remember, no soldiers in Cracow were allowed to leave their barracks. Nevertheless I decided to go to the lectures. A colonel stopped me on the way home from the university. I showed him my forged pass on which I had written with red ink *exceptional case*. The colonel asked me why I was allowed to go out. With a tragic face and a nervous voice I answered, trembling at the sight of the gold collar and sky-high rank:

"My mother is dangerously ill, sir, and I was allowed to see her."

The colonel was touched. He said:

"Head up, soldier. Your mother may be well again. Hurry to her quickly."

The nervous strain brought tears to my eyes, and the colonel, sympathizing with my suffering, honored me by patting my back.

Some time later, walking through Cracow's principal park on my way to the university and pondering over uniformly continuous functions, I suddenly heard:

"You, soldier, come here!"

A corporal with a gypsy face and a Kaiser Wilhelm mustache stood before me.

"Why didn't you salute me?"

"I report obediently that I did not see you, sir."

"Oh yes. I knew that you would say that." Then, raising his voice: "When were you made blind? At the front?" I did not answer. He raised his voice higher. "I ask you, have you been at the front?"

"No sir."

"But I have been. See! I have been at the front. And you dare not to greet your superior who was at the front. You think that you can get by because you have an intelligence sign." The voice went higher and higher. "When you are a lieutenant I will salute you, but now you must salute me. See!"

My fingers bent automatically. I longed to put them around his neck with its two stars, to strangle him slowly, very slowly, to see the life leaving his body breath by breath. But at the same time I knew that once he denounced me to the commandant of my battalion my freedom would be gone.

Ashamed of my own cowardice, detesting myself, I said:

"I apologize, sir."

But this was not enough. The corporal would forgive me only if I passed him in six military steps and saluted him smartly. I did it and rushed home humiliated.

It was spring. The academic year was nearly over. I had not cut even one lecture. Samson systematically brought me passes and the sergeant received his weekly payments. How long could this last? One day Samson came and, after I opened the door, burst out without any introduction:

"I have bad news for you."

"What happened?"

"The sergeant was transferred from the office. There is a new sergeant now. He is a son of a whore."

This was a blow. I asked Samson:

"What do you advise me to do?"

He began to philosophize.

"If you report to the barracks, they will ask you where you came from. You will be in a jam. You could come and stay in the barracks so that nobody would notice you. But in this case, why go? You may just as well stay here. It would be best for you to show yourself piece by piece; very slowly, so that they would get used to your face gradually. But this is impossible. A soldier is an entity."

I was furious that Samson should be amused by my tragic situation. Apologetically he took out a bundle of passes.

"I am afraid that this is my last present. You have no choice. You may as well stay here."

A week later Samson came again, this time with the corporal who had so helpfully provided me with passes. In civil life the corporal was a Jewish scholar; though thin and weak, he had fought bravely in the Austrian army, had been wounded, rendered unfit for service at the front and was the brain of the fifth company's orderly room. Samson's and the corporal's faces were grave. They told me their story quickly. A checking commission from the battalion had come suddenly to the barracks in the middle of the night to verify the presence of all soldiers. A week before, an order from the battalion had canceled all permissions to sleep outside the barracks. The soldiers were awakened and assembled in the court. Luckily for me, the corporal was designated to read the names aloud. When he came to mine he omitted it and later added a check mark. But he could no longer help me. The barracks were responding to the increasing disorder in the country by increasing the military pressure. I must find some way out of the mess and take all the responsibility myself.

When they saw how seriously I took the news they tried to console me. A new checkup was unlikely during the next few weeks. I should have plenty of time to plan a way out.

My nerves began to crack. During sleepless nights I thought for hours how to get out of my situation. I became unstable, irritable, on the verge of a nervous collapse. The slightest provocation caused fits of temper in which I broke glasses and used the same oaths which had so disgusted me in the barracks. My grandfather threatened:

"I will go to the company and say that you are a deserter. You think that I am a fool and I don't know where your company is," and he described exactly how to reach our barracks. I was astonished. I had never thought that he knew the town outside the ghetto. Anything could be expected from my grandfather. I answered him as quietly as I could:

"You go. But before I go to jail I will do something to you. I would not like to be in your skin."

Even he became afraid of me. He sneered:

"With God's help I will go to your funeral."

I made my parents miserable by the sudden and unexpected outbursts of my temper. At night I was haunted by fears that I should end my life in a lunatic asylum.

My friendship with Samson became strained. He was always without money, and I had to help him. I was more interested in obtaining the supplies of passes from Samson than in convincing him that life was worth living. I am sure that in my thoughts I never connected the pass aspect with the money aspect of our relationship. But the fact was that practically whenever we saw each other I asked Samson for the supply of passes and he asked me for money. It was Samson who had the courage to say:

"Our whole relationship is deteriorating." He paused for a while and then added nervously: "You treat me like the sergeant whom you paid weekly, only with the difference that the price which you pay me is irregular and not fixed."

Just now, I thought, when the whole world is against me, he is starting a new quarrel. Unable to see his side, I answered angrily:

"Does it mean that you would prefer a fixed amount?"

Samson took the cheap Austrian paper money which I had just given him out of his pocket, threw it on the table and ran out without answering.

Of course I regretted my outburst. The next day Samson came as though nothing had happened between us, read Villon's *Testament* aloud to me, talked about Freneh literature and asked for money.

I was degenerating rapidly to the level where hate reigns. I

burned with hatred for the stupid faces of the army officers. I
wanted to annihilate the whole rotten Austrian army machinery.
For hours I daydreamed of meeting the same officers who had
humiliated me and of taking a fearful revenge. I imagined myself talking to them in the same language they had used to me,
beating them brutally and spitting in their faces. Breathing the
foul war air, even behind the front, was enough to make one's
mind sick.

Everyone hated the war. The women were more courageous
than the men. After three years of patient standing in queues,
they broke shopwindows and threw stones at policemen. Nobody knew what was the purpose of the war. As long as old
Emperor Franz Joseph I lived we soldiers wore the letters F.J.I.
on our caps. The standing joke was that the initials represented
the war aims: "For Jewish Interests." *(Für Jüdische Interessen.)*

I nursed the same naïve thoughts as millions of others. Why
did the war go on? If one soldier at the front would say "to
hell with the war" and go home, everyone, absolutely everyone,
would follow him because everyone felt as I did. Even the officers had had enough. They were humiliated and pushed around
by superiors just as we soldiers were pushed around and humiliated by corporals. They humiliated and suppressed us in return
to rid themselves of the accumulated bitterness. It was a strange
pyramid, with ordinary soldiers at the bottom, bearing the greatest pressure, and with a few men at the top. Who were these
men? Who were the few criminals who prolonged the war and
forced the fighting to continue without sense or reason, uttering
meaningless sentences about the future glory of the Austrian
and German empires?

The restless atmosphere in Cracow increased. Hungarian soldiers were moved toward Cracow and Polish soldiers toward
the Ukraine. It was simpler to keep the empire intact by letting
the Hungarian soldiers shoot the Polish civil population and the
Polish soldiers shoot the Ukrainians.

I hated the war but still more I hated my precarious situation.
I held a council of war with Samson and the good Jewish corporal. My war plans were complicated and dangerous, but I con-

vinced them that they must do something to save my skin and theirs.

They agreed to try. They were to give me papers transferring me to the convalescent home so that I should cease to exist in the books of the company. How I should convince the doctors in the convalescent home that I was sick was my problem and only mine. The corporal was to obtain, in some miraculous way, the signature of the medical officer which I should present at the convalescent home.

Before the plan was completed I had to find out for what disease I was sent. I went to my private doctor who found that my heart was much better and, if investigated, would lead me straight to the front. Men were needed more than ever. It was only the strain which had caused the arhythmic beating before and which, luckily for me, had been mistaken for an organic illness.

The Austrian machinery was great, disorderly, corrupt through and through, becoming worse with each day that the war progressed. Single-handed I had to fight this monstrous machinery, to find my way through the maze of its rusty wheels, to steer cautiously toward freedom. Sometimes I think that no scientific work which I have since done required as much concentration as the problem of leaving the Austrian army.

How to convince a score of fools whose language I didn't speak that the best thing for them would be to leave me alone? My body was healthy. But does a healthy body always contain a healthy mind? What about simulating insanity? It is not simple. "Crazy" people are fed castor oil, starved and humiliated until they become crazy or confess. But I could be unbalanced, explosive, stirring up fights with everyone, near to a nervous breakdown, a danger to the morale of the Austrian army. Imagine Samson's restlessness and nervousness developed far, far beyond his present state. Samson to the third power would be impossible in the Austrian army. How would such a type behave? He would constantly wrinkle and unwrinkle his forehead, close and open his eyes frequently and irregularly; he would have nervous tics; he would talk not in sentences but in disconnected words;

his fingers would be in constant motion, his eyes staring rigidly in one direction and then flitting in a disorderly dance around the room. He would bang his foot on the floor; his thoughts would take sudden, unexpected turns; he would mumble the most unconventional phrases.

I thought out this type carefully and studied it before the mirror. I did not know one essential fact: that what I studied was not Samson's type but my own. What I really did was to study the symptoms which were in me, trying not to hide but to expose and magnify them. I practiced the type on my parents and with special pleasure on my grandfather, exploding with anger, cursing and frightening my family. Later, ashamed of myself, I tried to analyze how great was the genuine component and to what degree it was an intentional performance. I never found out. My confusion grew; I did not know when I was acting and when I was revealing my own sickness.

The plan worked well in the convalescent home. From there I was sent for investigation to the psychiatric department of the garrison hospital.

The first doctor examined me carefully for nearly an hour. He checked my reflexes, flashed a light in my eyes, made red marks on my skin, looked at my outstretched fingers and asked me to show him my tongue, at which he stared for a while. Then he asked me about my symptoms, how I felt, what kind of fears I had, and finally fired an unexpected remark:

"I suppose you intend to work scientifically in the future?"

"Yes, I do."

"That will be all. You may dress."

But one doctor, and especially an intelligent doctor, did not mean much in the Austrian army. Often at the bottom of the hospital ladder were intelligent reserve doctors with splendid civilian practices, forced to serve in the army for the duration of the war. But higher on the hospital ladder were army doctors with fine gold collars, born in army uniforms and intending to wear them to the last day of their lives. The procedure demanded that I be examined in succession by four doctors. First the civilian doctor. Then I would go to the army doctor who had the right to change his predecessor's diagnosis. Then I would go to the army doctor who was the head of the psychiatric department. There I should meet a gold collar with the right to change everything that the two previous doctors had done. Then would come the worst ordeal. I would have a brief interview with the commandant of the whole hospital, a major general, who would either sign the final certificate or change it completely after a one-minute interview. He never examined anybody (I wondered: did he know where to look for a pulse?). He simply took one look and either signed, tore up or changed the final diagnosis.

My turn for the interview came. The gold collar turned to me and said in a sharp military tone:

"You are a malingerer. You ought to go to the front."

This was exactly the remark I expected, and my answer, in its content and delivery, was carefully prepared. Without using the "sir," "Major General" or "obediently," I mumbled in a civilian way:

"I am happy that you said it. I ought to be at the front. I know it. Why have other soldiers to fight for the fatherland and not me? But why do you say that I am a malingerer? You have no right to say this. I did not want to come here. They sent me here. Why do you say that I am a malingerer?"

I began to repeat the last sentences over and over again when I heard the simple command:

"Shut up."

I still mumbled the same words in a lower tone until I heard in a much louder voice:

"Shut up, I said."

The gold collar looked at me as though he would like to spit in my face, changed something in the report, signed it and dismissed me.

The diagnosis was: *"Geistig minderwertig."* I was feeble-minded. My mental level was depreciated below the level required of the Austrian soldier. One of the symptoms, according to the report, was that I had a smooth tongue, without lines. The name for it, written out in brackets, was *"Idiotenzunge,"* the tongue of an idiot. My whole life I had kept a treasure in my mouth without knowing it. A bad tongue is much better than a bad heart. It would not disappoint me, it would remain smooth; to show it around would be sufficient to convince anyone that I was unfit to serve in the emperor's army.

The procedure was by no means finished. With this hospital report I was sent to the company doctor, who looked at my tongue, then to a great commission of officers and doctors to whom I again stuck out my smooth tongue; and the final decision was that I should get a leave for three months, after which I should have to return to the hospital to see whether my mental state had reached the required level.

I dressed in my old civilian clothes. It was a good feeling. No saluting for me! On my first day of freedom I entered a streetcar and saw an army officer brutally pushing and elbowing his way through the crowded car, stepping arrogantly on civilian feet. Suddenly I heard my own voice speaking sharply:

"You officer, behave properly. You think that you are in your barracks where human beings are treated like swine. I know how they are treated in the barracks. I have been in the barracks. But we are not your slaves here."

The sympathy of the car was with me. But the officer personified the power and dignity of the Austrian army, and no one upheld me openly.

"Stop the car until I come back, and don't let anybody out," the officer commanded the motorman. He soon came back with a policeman and charged me with insult to an Austrian uniform. It was not he whom I offended but it was the uniform which I dared to insult.

There was a soldier in the street car. The officer asked him:

"Did you hear what this civilian said to me?"

"Yes sir, I report obediently I did."

"Then you go as a witness."

We were both taken before the magistrate. He was in civilian clothes, and everyone in civilian clothes seemed to me a harmless angel of peace. The magistrate listened to the policeman, then to the soldier, who suddenly lost his memory and said that there was some loud talking but he did not know what it was about.

The magistrate turned toward me:

"We are at war. You had better keep quiet, young man. I shall have to punish you," and he mentioned a ridiculously small fine, which I paid.

The three months of freedom were gained during my second year at the university. Professor Smoluchowski died suddenly, and I lost one of the greatest teachers I ever had.

I began my studies of theoretical physics with mechanics and later read books about Maxwell's theory of electricity. It was a most impressive logical structure. From a few simple assumptions a whole theory was built up, perfect in its simplicity and mathematical beauty. A theory which not only explained an immense range of known phenomena but led to great new discoveries. It was by Maxwell's theory that the existence of electromagnetic waves was predicted before they were discovered experimentally by Hertz.

Now, in my second year of study, I plunged eagerly into theoretical physics, in which I saw the synthesis of mathematical beauty and the vividness and attraction of reality. A theoretical physicist does not need to deal with messy experiments. He attempts to penetrate by thought deeply into an understanding of our world; he tries to enclose the laws governing the universe in simple formulae; he tries to devise a logical system representing the law and order of the outside world.

The professor lecturing on theoretical mechanics ended his course by devoting a few hours to Einstein's special relativity theory. For the first time I heard the name Einstein, for the first time I heard about the Lorentz transformation which Einstein had formulated. It was a revelation. The beauty of this structure, the courage to follow an entirely new point of view to seemingly shocking results, revealed the scientific imagination of a genius. I was not sufficiently prepared to understand the entire structure of relativity, but I knew that I should return to this theory in my later studies.

When, some months later, I reached an understanding of

Maxwell's equations an idea occurred to me: to apply the
Lorentz transformation to electromagnetic phenomena and to
see whether Maxwell's equations are invariant with respect to
this transformation. During my third year of studies I thought I
had found something new and important. I rushed to the professor to show him my result, only to learn that the work had
been done far better, in a much more general and complete
form, thirteen years before by Einstein and Poincaré, that it
was in fact this problem from which the whole relativity theory
arose historically and that the investigation was later perfected
by Minkowski, who had developed the appropriate mathematical technique in 1908. It was a good lesson for me. I saw that
there was still a great deal that I had to learn before I attempted
original work. But the excitement of formulating a problem, of
solving it for myself through constant thought and sleepless
nights, gave me the first taste of the joy and suffering which
are a part of all scientific work.

    After the three months of my military leave had passed the
leave was prolonged for a further six months. In obtaining them
I made good use of the malingerer technique which I had
learned. My nervous state improved. Before the six months were
over the war ended and Austria ceased to exist. Poland was
created, and French and American pictures were shown in the
movie houses. Wilson was proclaimed the greatest man in the
world, and I learned that God had always been on the side of
the Allies. Everyone performed an about face. From Austrian
patriotism, through the indifference of the last years, to a pro-
Ally attitude so maudlin that even the French were embarrassed.
If I had delivered a speech in the Cracow market place recount-
ing all my forgeries and lies in the Austrian army, I would
have been hailed as a jolly good fellow for what would have
led to years of imprisonment two weeks before.

    I accompanied Samson to the train which took him to Vienna,
now in another country. We were to be divided by new
frontiers, and God knows when we should see each other again.
I knew that he would never write.

    I went home to my new surroundings, and with perverse

pleasure I used my soldier's coat as a smoking jacket. I had started my service as an ordinary soldier and finished it as an ordinary soldier.

Peace returned for a brief stay to Cracow. With hope and curiosity I looked into the future.

VI

IN THE FIGHT FOR POWER in young Poland the reactionary forces won. Soon I found out what anti-Semitism meant. General Josef Haller, who had organized a Polish legion in France, came to Poland. His soldiers amused themselves by catching orthodox Jews in the streets of Cracow and tearing out their long beards. The more gentlemanly soldiers used knives. The reactionary Polish press regarded this as a good joke. The excuse was that the soldiers were very civilized and therefore annoyed by the barbarous Jewish beards; they wanted to give a civilized appearance to the Polish towns. Some orthodox Jews with a bitter sense of humor carried emergency scissors in the hope that the attacking soldier might be persuaded to use scissors instead of fists.

Anyone reading the Polish newspapers at that time would have thought that the new country had but one burning problem: What to do with the three and a half million Jews? Collected almost exclusively in cities and towns, they comprised one tenth of Poland's population. In Cracow Jews were accused of being Austrian rather than Polish. In Warsaw Jews were accused of being Russian rather than Polish. In all Poland Jews were accused of being Bolsheviks. A pogrom broke out in Lwow in which hundreds of Jews were killed. Again the ultranationalistic press excused it by claiming that Jews had poured hot water from the windows of their ghetto on marching Polish soldiers. Words of protest and condemnation were weak and isolated.

To justify anti-Semitism the Polish nationalistic parties claimed: "Jews are our worst enemy." It is a simple and successful method, and it works beautifully all over the world. The first step is to put this slogan in many mouths, hearts and minds. The second step is to treat Jews badly, to treat them as enemies. Then the Jews must fight back if they have a chance to do so. They would be inhuman if they answered hate with love and devotion. In this way the Jews themselves must sooner or later provide the justification for the first step. It is the last link. The proof that Jews are a bad lot is given. The ring is closed and the picture is complete.

It is difficult to understand the logic and arguments by which anti-Semitism works. Sometimes the argument runs like this:

"The Jews poison our culture. We must keep our literature, our books, schools, universities, theaters, free of Jews. They bring decay and disintegration wherever they enter. Let them— at the most—be allowed to publish books in their own language, to teach in their own schools but never in ours. Jews who cannot leave the country must go back to the ghetto where they belong."

Again the thesis is reversed:

"Jews, closed in their ghetto, form a strange and hostile body in our national organism. They are parasites on our economic and cultural life. They reject our language which they don't speak after having eaten bread for centuries from Polish soil. They will never assimilate; they want their ghetto. Come out of the ghetto or you are our enemies."

The Polish press emphasized the latter argument, throwing it thousands of times in the faces of the Jews. The assimilationists, the Jewish intelligentsia, who had been quite numerous when Poland was a suppressed nation, lost all their influence among the Jewish population. Thus anti-Semitism in Poland achieved the opposite effect of its supposed aims. Jewish national feelings soared and, at the same time, were blamed as the essential reason for anti-Semitism. The atmosphere of conflict and hatred pushed the Jews further into their isolation. Zionism, never before strong among Polish Jews, increased rapidly. I myself went through a short period in which, under the pressure of anti-Semitism, I believed this to be the only possible solution of the Jewish problem.

Always before, Poland had been for me a heroic, unjustly suffering country, and its fate, legends, literature, expressed in the powerful form of its poetry had made my cheeks burn and my eyes fill. The picture of Poland, suffering and oppressed, was inconsistent to me with the picture of Poland oppressor and author of suffering. The rise of my Jewish nationalistic feeling was a self-defense against disappointment. It was a new outside force, crowding me once more toward the ghetto.

Jews began to fight back. They wanted a "Jewish autonomy," the right to have their own schools, the right to be treated as a respected minority, to preserve and to develop their own culture. Owing to their influence in Paris, a minority clause was inserted in the Versailles Treaty which, on paper at least, gave them the rights for which they fought.

In Polish towns Jewish gymnasia grew suddenly, like mushrooms after a rainy day. Jewish communities, social organizations, parents' committees, citizens' committees, private individuals, began to open Jewish gymnasia. Even plain businessmen did it because it turned out to be good business to squeeze the parents for as much as possible and to pay the teachers as little as possible. Buildings were acquired for money, children provided by the Jewish ghetto, which was ready to send its sons and daughters to Jewish gymnasia. The greatest difficulty was the lack of teachers. There were not enough of them, by far not enough. Disappointed lawyers who could not make a living taught history or Polish literature; engineers who could not find places in industry or polytechnic students who could not go on with their studies taught physics and mathematics; young medical students without means to continue their studies taught biology.

The "philosophical studies," previously regarded as "Gentile studies," suddenly became crowded with Jewish students. The Jewish gymnasia created new opportunities. Bad and dilettante as they were, they nevertheless performed a great service. They brought education to parts of the ghetto which had been immune to the influence of purely Polish schools.

I was in the third year of my university studies when the gold rush to the Jewish schools began. Without any diploma, I could have interrupted my studies and started teaching. But I enjoyed my studies too much for that, and I was sure that the rush would endure for some time. In two or three years, when I finished, I would still be one of the rare birds with a real teaching diploma.

But how different was the reality from the dreams of my youth! Now to teach in a gymnasium was almost a disgrace. Men who had failed in other occupations took up teaching. The Jewish gymnasia were caricatures of my old idealized picture. Reality made my cherished dreams vulgar and cheap.

Although very good in mathematics, the university in Cracow did not offer much in theoretical physics, the subject which attracted me most. The only lecturer in mathematical physics was an old, completely detached professor, delighted with the smoothness and external beauty of his lectures and not really giving a damn whether he inspired anyone or not. For thirty years he had lectured in Cracow and had never had a Ph.D. student. But to do scientific work one must learn its technique. Only a genius can do without that. I well knew that if I were to amount to anything I should have to learn and learn. I wanted to go to a university with a good reputation in theoretical physics and to work for my Ph.D. But physics was not my only incentive to leave Cracow. I had never set my feet more than one hundred miles outside my home town. I wanted to see the world outside the little corner in which I lived. I was still immersed in the ghetto. There I stayed with my parents and there I had my friends. I created a protective layer around me by studies and books. By living in an atmosphere of abstract scientific problems I tried to diminish the impact of my own environment and the impact of social problems carried to my world on the waves of anti-Semitism. But I became more and more weary of the constricted life around me and the dullness of playing with the same words and ideas.

I became weary of the political atmosphere in Poland, of the growing anti-Semitism and the tensions which were gradually diffused into the university. At this time there was war between Poland and communist Russia. The Ukrainian army, under the leadership of Petlura, fought hand in hand with Poland. Strange rumors penetrated into Cracow: News of pogroms, towns burned, mass execution of Jews, systematically and methodically performed by Petlura's men in Ukrainia's towns full of Jews. They sounded incredible and made the Polish anti-Semites look like innocent angels by comparison. (Later their truth was demonstrated to the whole world when a Jew who killed Petlura in Paris was acquitted by the French courts.) Under the pressure of war psychosis even the Socialist party in Poland defended the necessity of war with Russia to liberate the Ukraine, to create a block of independent countries between Poland and the Soviets. But the same Socialist party which took the view that the Ukraine must be liberated, since only under a Polish protectorate could it develop its own political and cultural life, declared that Jews could be digested only if they assimilated to Poland. Thus even the Socialist party conducted a mildly aggressive external policy and a mildly aggressive anti-Semitic internal policy.

To go away one had to have money. I talked to my father. He agreed immediately to keep me in Germany for a year. I could guess why. The war atmosphere was growing, the Polish army formed, and he was afraid that my turn would come. He sensed that I could not use the old Austrian tricks in the new Polish army. Poland, right or wrong, was my country. He thought that the safest thing was to keep me outside Poland. So I went to Germany, intending to study at the University of Berlin, at that time one of the best universities of Europe.

With wide-open and astonished eyes I tried to absorb the wonders of this great city with its cosmopolitan atmosphere, its fine libraries, theaters, a great university, a town in which life was wonderfully organized. Order, order everywhere. Attention! Watch your step! Go right, go left! The touching of cakes is forbidden by the police! Do this, don't do that! Still a country of order and the fear of God in spite of the lost war.

"Here I will study. Here Einstein lives, the greatest physicist of our generation. Here Planck lectures, who first formulated the quantum theory, and Laue, who discovered the diffraction of X rays by crystals, dozens of lecturers in theoretical physics alone. The first thing is to join the university and plunge deeply into this new, strong atmosphere of research."

But it was not so simple. If there was a nation hated more than any other at this time in Germany, it was Poland. Among the wounds inflicted by the Versailles Treaty those of Danzig and the Polish Corridor seemed to be the most painful. And they were inflicted by a nation which Germany had previously conquered and, in their conceit, had thought to be weak and worthless. In Poland I was a Jew and not a Pole. But here in Germany I was a Pole, a member of a hostile nation. I learned that it was impossible for a Pole to be admitted to the university without powerful outside influence. I tried to achieve this through Jewish channels. Some influential people of the Jewish community in Cracow gave me letters of recommendation to influential people of the Jewish community in Berlin. Wherever I went I was received with a tremendous air of superiority. They let me wait in the hall, then, smoking their big cigars, put on their glasses and glanced at my papers.

"So you passed the Polish matura and you studied at the Polish university in Cracow? I am afraid it is not of much value here. Why didn't you continue in Cracow?"

In these interviews I learned about the superior attitude of German Jews to any other Jews in the world and especially to Polish Jews. Among Polish Jews, in turn, those from the Austrian part of Poland, from Galicia, were regarded as most inferior. To the Germans I was a Pole who had grabbed Danzig and the Polish Pomorze. But to the German Jews, enjoying the blessings of the superior German culture which spread order and obedience everywhere, I was an "Eastern Jew," lowering by my appearance the high level of their lives and thoughts.

One of the men who interviewed me said:

"If you are a physicist, why don't you go to Einstein? Maybe he will help you."

I thought it impudent to bother Einstein with my troubles, but it was the only thing which remained. I rang up Einstein's home:

"Is Professor Einstein there?"

"Yes, he is," a woman's voice answered.

"I am a student of physics from Poland and I would like to see Professor Einstein. Could Professor Einstein grant me an interview?"

"Certainly. The best thing will be for you to come right now."

Shy, deeply touched, in a holiday spirit of expectation at meeting the greatest living physicist, I pressed the bell of Einstein's flat at Haberlandstrasse 5. I was shown into a waiting room full of heavy furniture and explained to Mrs Einstein why I had come. She apologized and explained that I would have to wait because a Chinese minister of education was just then talking to her husband. I waited, my cheeks burning with excitement. A few minutes later a young man with a thin vivid face and smiling eyes entered the room and sat down opposite me. He put the book which he had brought with him on his knees; it was Weyl's *Space—Time—Matter*, a book which became a classic in advanced relativity theory and which I was studying carefully at this time. We entered into conversation. I asked him:

"Are you a physicist?"

"No. I am a philosopher; I recently took my Ph.D. in Prague, but I am studying now at the University of Berlin and I am interested in the philosophical aspects of the general relativity theory. There are some things in the last chapter of this book which I don't understand. I should like to discuss them with Einstein."

I was astonished at the idea of coming to Einstein so freely. I asked:

"Do you know Einstein?"

"Yes, I know him from the time he was a professor in Prague. My father was on the same faculty, and they knew each other quite well."

Our conversation was interrupted by Einstein, who opened the door of his study to let the Chinese gentleman out and me in. Einstein was dressed in a morning coat and striped trousers with one important button missing. It was the familiar face which one saw at that time so often in pictures and magazines. But no picture could reproduce the shining glow of his eyes.

I completely forgot my carefully prepared speech. Einstein looked at me with a smile and offered me a cigarette. It was the first friendly smile directed toward me since I had come to Berlin. Briefly I told him my situation. Einstein listened carefully. There was none of the impatience and desire to finish the interview which I had encountered everywhere else.

"I should be very glad to give you a recommendation to the Ministry of Education. But my signature does not mean anything."

"Why?"

"Because I have given very many recommendations, and"—here he lowered his voice to a confidential tone—"they are anti-Semites."

"Then what would you advise me to do?"

He thought awhile, walking up and down.

"The fact that you are a physicist makes it simpler. I will write a few words to Professor Planck; his recommendation may mean much more. Yes, this will be the best thing."

He began to search for his writing paper which was on his desk before him. I was too shy to point it out. Finally he found it and wrote a few words.

By this time I felt so free with Einstein that I asked him a question connected with relativity theory: what he thought about Weyl's new generalization.

"No, I don't like Weyl's new theory, but I like his book very much. You see, if you take two hydrogen atoms and transport them from the earth to the sun along two different paths, they will have different frequencies according to Weyl. I don't believe that the frequency of the atom depends on its past." And then, laughing aloud, he added in a childish tone, "No, I don't believe that."

Then I asked him about the momentum energy tensor in relativity theory. What does it really mean?

"This is difficult to answer. I say in my lectures that the general relativity theory rests on two pillars. One is very beautiful and strong, like marble. This is the curvature tensor. The second is weak, like straw. This is the momentum energy tensor." Again the same broad laugh. "We must leave this problem to the future."

I remembered that the philosopher from Prague was waiting and did not ask any more questions.

This was my first and, for the next sixteen years, my only personal contact with Einstein. At our first meeting in Berlin (and many times later) I found the old platitude confirmed, that human kindness and real greatness are invariably found together.

VII

I wrote a long and detailed application to the Prussian Ministry of Education, asking for permission to study at the University of Berlin. It was the summer of 1920. While I waited for an answer I worked during the day in the great state library, eating cheap, indigestible food and saving money for standing room in the theaters to see Moissi in *Romeo and Juliet*, Basserman in *Othello*, Wegener in the *Merchant of Venice*. My only acquaintances were a small colony of Polish students, most of whom were waiting, like myself, for admission to the university.

At last the important day came. The list of the admitted students appeared. I took a quick glance at the list, nervously running through the names. I did not see mine. I looked again, trying to calm myself and to read it carefully. My name was still missing. Perhaps it was a mistake. I went to the office and asked. No, it was not a mistake. My application was rejected.

The weather was fine, the sun shone, the houses stood as before, the streets were gay and the whole town conducted its business as usual; nobody took any notice of my suffering. I went to the subway, toying with the idea of making a misstep and imagining how the rushing car would impersonally and completely end the accumulated pain. As in a dream I went to my room, lay down on the sofa and idly picked up one of the books the landlady had left in my room. It was a volume of Heine's poems. I read them aloud one after another, not understanding their meaning but hearing in their musical rhythm the reflection of the suffering world in which my suffering was the burning center. I put aside the book, went to the nearest railway station, took the next train and got off at the last station. It was Potsdam. I went through the garden and palaces in which Friedrich the Great and Kaiser Wilhelm had lived and listened to the guide's mechanical explanations. The mixture of luxury and bad taste sickened me. On the way home in the evening I went to a beer garden and drank one heavy beer after another. Exhausted and half drunk, I went home and slept. Next day I felt the strength to face my problems. "There must be some way out," I repeated to myself. "It would be silly to give up everything now and return to Cracow." The idea of returning frightened me.

I remembered that among all the Polish students only one had been accepted. How did it happen? He was younger than I and, as far as I knew, had no recommendation. In spite of his youth he was, in practical everyday problems, much more mature than I; one of the type born in long trousers, with eyeglasses and a perfect hair-do, who remains static all his life.

I looked him up and he told me his story:

"It is a scream. You won't believe me. But here it is. I have a cousin here. He is terribly rich. He made big money during the war and after. No education, rough and shrewd, the kind of guy who knows how to get things going. He has a mistress. You ought to see her. Oh boy! Has she a figure and a skin! This mistress has a husband who is a teacher in a gymnasium. All three of them go to night clubs together."

When I opened my eyes with astonishment he explained to me:

"You don't know Berlin. It is one great brothel. You can't imagine what is going on here. The teachers are starving, and this man is lucky to have a wife for sale. Anyhow, they took me to a night club. The teacher was drunk and told me that he could easily help me get into the university. He wrote a letter, 'Dear *Obergeheimrat* R: This man is my relative and it would be very nice if you could help him.' The whole thing seemed screwy to me, but I thought that I hadn't much to lose and I enclosed the letter. And imagine! It worked. A few days ago the teacher showed me a letter from *Obergeheimrat* R. I have seen it. There it stood. 'It is a pleasure for me to do something for my old schoolmate.' Yesterday I bought a fat goose and presented it with my compliments to the teacher and his wife. So you see, it was simple. I was astonished myself."

I believed the story but hardly knew what to do. I again began the familiar round, starting with visits to Jewish dignitaries. This time I was less coldly received. It was no longer merely a problem of a Polish Jew. They knew that Einstein had written a letter on my behalf to a German professor, who had recommended me because of the letter from Einstein. This meant that Einstein, the pride of world Jewry, had been ignored by the Prussian administration. I finally landed at the office of Mr Cohn, a member of the Prussian Parliament. He gave me a brief interview, was very matter of fact and asked me to leave the papers and my address. A few weeks later I obtained from the Ministry of Education a post card stating that I was to be allowed to enroll as a "special student." Although the time did not count formally toward a degree, it was at least some consolation that I could attend lectures.

Professor Laue lectured at this time on relativity theory, and I attended his course. Here I again saw Joseph, whom I had met when waiting for an interview with Einstein. We talked before and after lectures, walked and lunched together, visited each other at home and worked together until our friendship was cemented. It has proved its strength up to this day, in spite of our geographical separation.

Later Joseph and I organized a small group in which we discussed carefully the more difficult and subtle points of relativity theory. One of the members of our circle was Grommer, who had published a few papers with Einstein, a phantom whom everyone in Berlin knew by sight. He had a rare illness called acromegaly. His bones had grown so enormous that his chin was bigger than my whole face. When I shook hands with him my palm was drowned in an ocean of flesh. The illness deformed his body and affected his mind so that later he became sterile and his scientific work ceased. But at this time he was still in good form and in a coarse voice, mixing German, Russian, Yiddish, discussed with us the subtle mathematical problems of relativity theory. Until late at night we talked about the meaning of measurements, the rotating disk, curved light rays, and analyzed the fundamental assumptions on which relativity theory rests.

But more often only Joseph and I met, and our discussions ranged from relativity theory to philosophy, from personal confessions to social problems of which I had been little aware in Poland. But in postwar Germany, in the hot atmosphere of class struggle and bitterly divided parties, the social problems could not be ignored. They cried loudly from streets covered with slogans, from daily papers, from bookshop windows, from the tense atmosphere of the city created by the mixture of poverty, unemployment, luxury, night clubs, prostitution and decadence.

Once I looked through Joseph's library where Marx's *Das Kapital* and Engel's books stood beside philosophic and physical treatises. I heard Joseph saying:

"I am a member of the K.P.D. Perhaps you don't know what the letters mean. It is the German Communist Party."

"Oh, yes, I know what those letters mean. I also know that in

Poland you would get eight years in prison for belonging to this party. Why don't you belong to the Socialist party if you wish to save mankind? It has the same aims and it is much more respectable."

Joseph smiled ironically.

"If you saw two shops both proclaiming 'we sell fresh eggs' and you found by experience that one of the shops sold bad eggs and the other fresh eggs, would you still claim that the two shops are alike because they have the same sign?"

"Is it as simple as that? Both parties grew from the same problem. Both parties regard Marx as their prophet. They both claim the same aims. Is it right to make such a fuss about a few details?"

Joseph never got excited. Adjusting his pince-nez, he explained in a scholarly, quiet way, smoothly, like a man who knows all the answers before the questions are formulated.

"Let us start with the aims. There is a fight between capital and labor. Where is your place in this fight?"

"Practically nowhere," I answered. "Up to now I have not done anything, and I don't know that I shall have the guts to do anything in the future. But emotionally I am on the side of labor; fight for the underdog, equal opportunities for everybody, exploitation of the working class—all these slogans are familiar to me and I respond to them properly. In other words, I have a social conscience. It is asleep but it is there. You may assume it and go further."

"But if you once assume this, then the question of creating a new, just world, of building it up, is not a technical detail. I could reverse your argument. Nearly everybody wants the world to be better or at least claims so. The only essential problem is *how* to achieve this end. And here we differ. Read history and you will see that the German Social Democratic is a bourgeois party deceiving the worker, selling him the bad eggs and doing everything in its power to prevent the customer from using the other shop."

He began to quote facts:

"The Socialist party in Germany voted for the war in 1914.

Liebknecht, the only man who had the courage to condemn the war, was fought and was murdered under the very nose of the Weimar Republic. The Social Democrat Noske sent Prussian troops to Munich to squeeze the communistic Bavarian regime. Kautsky claims that the worker has to fight only to keep what he has. He has enough. The Social Democratic party is dangerous because it deceives workers and kills the revolutionary spirit of the working class."

"Are you sure," I asked Joseph, "that there are no bad eggs in your shop? Let me quote some facts about my country. Poland started an adventurous and dangerous war against Russia, I grant you. Later the Red army repulsed Polish soldiers. But instead of stopping at the Polish frontier they went deep into Polish territory and besieged Warsaw. Peasants and workers who had been against the adventurous war before went to the defense of their country. And what happened? The defeat of the Russian army near Warsaw is now called the 'Miracle on the Vistula.' It will strengthen the reactionary forces in Poland for years to come. Can you defend the Russian war against Poland, deep in Polish territory?"

"It is easy to criticize now," Joseph answered, "after the war has been lost. But suppose that it had been a successful war. Then the revolution would spread to Poland and from there toward the west. It would have freed the Polish worker and peasant. It is not possible to form a communistic order in one country, even as big as Russia. The whole capitalistic world, including the Second International, is against Russia. Our aim is world revolution, and I believe our goal is near. Look at Germany. Misery, poverty, profiteering, decay, the unbearable pressure of the Versailles Treaty—all that must lead to an explosion. It will be a painful process. But we shall be in a good position to try the communist experiment. The success of communism in Germany will convince the worker everywhere. It is not like Russia, which was one of the most backward countries in the world, with slavery and without industry. The task of building a communist regime in Russia is so difficult that it will take years until the workers of the world will see the effects through the organized lies of the capitalistic press."

A thought occurred to me which made me laugh. Joseph looked at me, smiled too and said:

"It will be a world worth having."

"I must confess," I said, "that I smiled for different reasons. I pictured a revolution, barricades, fights, shooting, then terror —always necessary to keep up the revolutionary spirit and to frighten the enemies. What would you do and what could you do in such a world with your purely theoretical interests? I cannot imagine you fighting, killing or proclaiming death sentences. You would write a history of the revolution after everything is over. Is it not again the same problem? We can hardly understand the reactions of the worker. We really belong to the privileged in the capitalistic system. It is difficult to get rid of this psychology. And without knowing it you are apt to deceive the worker."

Joseph answered quietly.

"I am sure you are mistaken. I am sure that I would have gone out and fought if the barricades were in Berlin. I am sure I would. You are right that the intellectuals often deceive the workers. But not always. You have among intellectuals leaders like Marx, Engels, Lenin, Liebknecht, Rosa Luxemburg and traitors like Kautsky. Our party here is really run by workers."

And then, after a while, he added:

"All that I can tell you you will find in books." He turned toward his bookshelf, took down a few books, *Das Kapital* among them, and handed the bundle to me. With the exception of the story of Liebknecht's tragic life, I had never studied one of them. I was tremendously interested in talking around the subject, and we spent hours at it, but I was afraid of *Das Kapital* and too busy with my scientific studies.

One day I was invited to give a few popular lectures about relativity, which reached the peak of its fame in 1920–21. It was a private gathering where I met a few interesting people with whom I later discussed our "Weltanschauung" in the long evenings, quarreling, often using immature arguments, mixing them with emotions, employing the heavy German language to obscure simple meanings with long words and involved sentences.

Again social problems came up in our conversation. One of the most fascinating types whom I met there was Dr B., a very successful young lawyer. Dr B., small, bald, with the power and oratory of a Robespierre, was one of the leaders of K.A.P.D. (Komunistische Arbeiter Partei In Deutschland), which regarded the K.P.D. as a reactionary party built by leaders, ready to compromise, a party in which the workers had little to say. In short, as I understood it, K.A.P.D. accused the communists of committing the same sins of which they accused socialists. I enjoyed bringing Dr B. and Joseph together and listening to their long discussions, with B.'s thundering and violent oratory and Joseph's quiet tone and sharp sentences. But in the maze of historical details, accusation and counteraccusation I became confused and did not know which of the many shops in Germany—S.D.P., U.S.P., K.P.D. or K.A.P.D. or still others—really sold fresh eggs to the worker.

Two things did crystallize in me in this hot political temperature: I overcame my Jewish nationalistic feelings, nourished by the anti-Semitism in Poland. I realized that suppression and hate is directed not toward Jews alone. Secondly I understood the danger of social isolation in the ivory tower which scientists like to build around themselves. I understood that a scientist ignoring his social duties, and refusing to see the ties which bind him to society, may find himself a victim of forces whose existence he has ignored.

If somebody were to ask me, "What did you do with your awakened social consciousness?" I should be forced to answer: "Practically nothing." But at least I have a bad conscience about it, and if this is the only difference between me and the others, I still believe that it is an essential difference.

I spent eight months in Berlin. For eight years afterward I looked back on those days as the most intense of my life. It was a rich mosaic. The background was the great city with its splendid theaters and the deafening noise of its streets. There was the university where I worked hard, climbing slowly toward those heights of understanding and knowledge where original work begins. There was the rich social life, friends who taught me to see events and problems in a new perspective. The tense life of a great nation struggling for an unknown future. The real seeds of this future were as invisible to me as to the best German.

The time I spent as a special student in Berlin did not count toward a degree; the time which I had spent at the university in Cracow did not count fully toward a degree in Berlin. It was only a Polish university! It would have taken me three years more to qualify formally for a Ph.D. in Berlin. Meanwhile prices rose and my money was nearly gone. I had to return to Poland.

I went back to Cracow, to the same room from which I had started eight months before. Everything seemed to me more strange and foreign than ever before, since I had glimpsed the great world. It was difficult to come back and to resume the old life. It was time to leave my old home in Kazimierz and to start working and living.

VIII

I CAME BACK from Berlin with some ideas about my doctor's thesis which I knew were not very important. Since I had not been a matriculated student in Berlin I could not have real guidance from any of the lecturers. Back in Cracow there was no one to give me helpful advice. But scientific research has its tools and tricks which one must learn by hard, careful, persistent work. I could not learn the tools of research in the most simple and effective way—personal contact. Neither was I mature enough to advance important new ideas. My first work was merely average: a modest contribution to the problem of light waves in relativity theory.

At twenty-three I obtained my Ph.D. in Cracow as the first doctor in theoretical physics in free Poland.

Poland had five universities. Two of them, in Cracow and Lwow, were old universities built by independent Poland and preserved by Austria. But three others, in Warsaw, Poznan and Wilno, were erected when free Poland was recreated. There were not enough scientists to occupy all the new chairs. Gymnasium teachers with very modest scientific accomplishments were promoted to be university professors. Especially in theoretical physics was there a scarcity of lecturers. As far as I know, in neither of the two Polish universities, Cracow or Lwow, had a Ph.D. in theoretical physics ever before been given. But I didn't deceive myself for a moment that I could obtain a position at one of the five Polish universities. Even my professor never gave his only pupil any hope for a university career. After the solemn and pompous ceremony at which, in Cracow, the doctor's degree is bestowed, he kindly invited me to his home. I sat in a fine armchair in a large study full of books. He asked me:

"What are your plans now?"

I answered:

"The most important thing for me is to be able to do scientific work in the future."

He looked very much interested in the theoretical side of this problem and said very slowly, carefully weighing each word:

"Yes, scientific work gives the greatest happiness and the greatest unhappiness. It makes life colorful and brightens its dullness. It is the most wonderful thing to do research without any disturbances from the outside world." With a dignified gesture he put aside all the disturbances of the outside world and proceeded: "In Cambridge, England, where I was thirty-five years ago, I always envied the Fellows in the colleges who, free of any obligations, could devote their whole time to scientific work."

I tried to be more explicit:

"It is wonderful to be a Fellow in Cambridge. But even the modest position of an assistant at a university gives opportunity of work."

To which he answered:

"I permit myself to disagree with you on this problem. There is really nothing like a fellowship in the Cambridge colleges. You see, in Cambridge . . ." And he went on skillfully from there to the educational system of Great Britain, not allowing me to come back to the subject from which our conversation had started.

When I left I imagined him saying to himself:

"What an impertinent Jew. He practically suggested that I might do something for him. Is he so naïve or stupid as to expect me to make him my assistant?"

There was only one way open: the way to a Jewish gymnasium. The bleak reality deformed the once glorious dream. Now I could have it.

But even this was not as easy as I had thought. For example, in Cracow there was a large and wealthy Jewish gymnasium. By Jewish was meant that it was for Jewish children, that besides the subjects taught in non-Jewish gymnasia something like ten additional weekly hours were added for Judaistic studies. What I had in my childhood in two different portions, the Polish and the Jewish school, was here amalgamated into one. At the same time the Judaistic studies were modernized. I would gladly have taught mathematics and physics in such a school. Staying in Cracow, I would thus have lived in a university town, could have used the university library and tried to do some scientific work. But life is not as easy as that. The gymnasium was built and administered by Jewish nationalistic elements, which can be exactly as reactionary as the reactionary Gentiles. To obtain an appointment in such a school the teacher had to show the same nationalistic spirit as the founders of the school and as the parents of its students. Even if someone taught mathematics and physics, his ideology had still to be well defined and free from doubt; he had to believe that the Jewish nationalistic life, based on Hebrew culture, ought to be preserved; he had to represent the social attitude of the Jewish middle class. After Berlin a place in such a gymnasium was, for me, perhaps even more hermetically closed than in a Polish university.

Many of the Jewish gymnasia in Poland had become inaccessible to me for the same reason. To the Polish world I was a Jew. To the Polish Jews I was not sufficiently Jewish.

For the next eight years I was a schoolteacher. For three years I worked far from cities, in small Polish provincial towns, in which nearly all ten thousand inhabitants were Jews. The children whom I taught came mostly from a background that made my home seem very progressive. The circle was closed around me. After all my attempts to move away, away from Cracow's ghetto, I landed in a town which was one great ghetto.

I asked myself again and again:

"Is this the end? Is there nothing more in my life?"

No library, no bookshops, no one with whom to talk even elementary physics. Dependent on stupid, narrow, dull people who were the trustees of the school, constantly fighting with them for decent salaries, burdened with teaching, hating the streets, buildings and faces around me, I slowly lost all desire to open a scientific book and forgot what I had learned.

Full of suppressed ambition and unfulfilled desires, I began to give up. My moments of greatest rebellion usually came when I was forced to walk downstairs with a candle on a freezing night and cross the yard to reach the outhouse. During these walks I had visions of the rich university life in Berlin, of science progressing, of my existence outside the living, struggling world. Lost in a wilderness of cultural isolation, I longed for the excitement of scientific achievement.

The provincial school in which I taught collapsed like many others. The Jewish middle class became poorer, and the teachers, organized into a union, refused to work for nothing—at least as long as there were opportunities elsewhere.

I gathered all my courage and the little money which I had, decided not to take another provincial job and went to Warsaw. I knew that it was exceedingly difficult to get a job in one of the Jewish gymnasia of the Polish capital. The Teachers' Union had succeeded in its fight for tenure. Everyone wanted to teach in Warsaw. Thus only death or accident could create an opening for which every teacher waited. I was lucky. In one of the best and richest Jewish gymnasia for girls a teacher of physics became seriously ill after the school year had begun and had to retire. The job was offered to me. I was thrilled and happy to be again in a large town with libraries, theaters and fresh faces; to teach in a school much superior in equipment and staff to any in which I had taught before.

I was so determined to succeed as a teacher, so overburdened with lessons and had forgotten so much that in the first few years I could do no scientific work. During my first year in Warsaw I taught thirty-eight hours a week. Slowly I turned toward books, toward the old half-forgotten problems. It took me a long time to understand my own thesis. There were so many new things to learn. The whole of quantum mechanics was developed during my provincial sleep; new fundamental ideas had come into physics. Science had not waited for me.

At least I had access to the library, to the scientific journals, magazines and books. There was only one professor of theoretical physics in Warsaw; he had previously lectured in a Russian provincial university. The first step was, as I thought, to pay him a visit and to join the Polish Physical Society, which, according to its constitution, was open to everyone recommended by two members. I went to Professor B. and introduced myself:

"A few years ago I took my Ph.D. in theoretical physics in Cracow. I am now here in Warsaw, teaching in a gymnasium, and would like to do some research if possible."

"Yes, I see," was his encouraging reaction.

"I wonder whether I could attend the meetings of the Polish Physical Society?"

"Yes, you could; as far as I know, the meetings are free for any guests."

I tried to imply that I would like to join the society by a diplomatic remark:

"I wondered about it, because I am not a member of the society."

His drawling answer was:

"You may not be allowed to take part in the discussions. But I don't see any reason for your being refused at the meetings."

The sweet little discussion was finished. The old feeling of despair returned: "Will I always have to pay with bitterness and disappointment for any attempt to move the smallest step outside my world?"

I went to the Physical Society meetings. There I met two young physicists who signed my application, and I became a member of the Polish Physical Society. Gradually I began to do some scientific work, and seven years after I had received my Ph.D. I succeeded in publishing in German and French scientific journals.

I began my eighth teaching year, the fifth in Warsaw, when I was thirty years old. I had ceased to believe that scientific work would give me anything more than the excitement of achievement. It was a good feeling, but I was afraid that even this would end. Scientific work needs encouragement, the presence of people with whom problems can be discussed exhaustively. With the exception of my short stay in Berlin, when I was too immature for research, I had never lived in a scientifically stimulating atmosphere. But I fully sensed its importance. I felt clearly that if one is not a genius the complete lack of scientific contacts must kill all desire to work. There was one danger more. In the isolation of Warsaw it was most difficult to work on vital problems on which groups of physicists worked with methods they created and developed. Like the Jewish Torah, which was taught from mouth to mouth for generations before being written down, ideas in physics are discussed, presented at meetings, tried out and known to the inner circle of physicists working in great centers long before they are published in papers and books. My defense in this predicament was to pick up problems off the highways of science, on isolated pathways in which only a few scientists, scattered around the world, were interested. This keeping out of the chief routes, this working in loneliness, sooner or later leads to faulty perspective. I did not really believe that I should be able to continue for long any research in this atmosphere of isolation.

I thought of my prospects for the future. I was thirty. For five years I had not done any scientific work. And these are the best years in the life of any scientist, the years in which imagination reaches its peak. Those years were gone; they had left only the memory of a miserable life in a provincial town and of the struggle for opportunities in Warsaw. My enthusiasm for work, in which I found an escape from the dull routine of teaching, would soon wear out. The salaries in the Jewish schools, reflecting the economic level of the Jewish middle class, were declining sharply. The number of students was diminishing. I had achieved the maximum which I was able to achieve. The doors leading to a university were closed and would remain closed. The doors leading to a non-Jewish gymnasium were closed and would remain closed. I was a teacher in the greatest town in Poland, in one of the best Jewish schools. From here only one way was possible—that leading down. I must enter this way. It was inevitable that I would fall, together with the Jewish school system, together with the increasing poverty of the Jewish middle class living in the ghetto. To this fate—now at thirty—I was bound as strongly and hopelessly as in the years of my youth. There was no way upward. All ways led only down to routine, to a repetition of the old lessons, to dependence on the same environment from which I had wanted to escape all my life.

Each morning I walked through the streets leading from my home to the ghetto community where my school stood. Again, as in my youth, I lived on the periphery of the ghetto and each day passed down the same streets—Orla, Karmelicka, to Nowolipki where my school was, in the heart of the ghetto, crowded by three hundred thousand Jews. Each day, each year, I saw the same sad faces of poor Jews, smelled the same smell of dirt and poverty, taught the same things, heard and told the same jokes. And I lost my belief in the future; I lost hope that my life might change tomorrow.

BOOK TWO

Escape

I

I RECEIVED an announcement of the Polish Physical Society's biennial meeting. It contained two questions: Did I intend to be present and did I intend to give a ten-minute report on my work? The meeting would take place in October (1928) at the university town of Wilno. I had never been in Wilno before. The name of this town stirred emotions in Polish hearts; interwoven with Wilno's past are the greatest names in Polish history: Mickiewicz, Kosciuszko, Pilsudski. Centuries ago Wilno was the capital of Lithuania, once united with Poland. It was the town still marked on Lithuanian maps and books as their capital, only temporarily under Polish occupation.

I wanted to go to Wilno and yet I dreaded to go to Wilno. I dreaded the hostile atmosphere of the university toward a Jewish teacher in a Jewish school. If I could only be sure of one human being with whom I could exchange a few words!

In the school where I taught there was an old teacher of mathematics, a kind old soul, bored to death by teaching and easily diverted. I asked him whether he intended to go to Wilno. No, he didn't. He was not interested in scientific work, he was not interested in sight-seeing and he was not interested in displays of snobbery. I pointed out to him the advantage of a week's absence from the school, how it would impress the administration that he had gone to a scientific meeting and, as a last inducement, how pretty the Wilno girls were said to be. If we were together we need not care about the others. We could amuse ourselves and watch the show. He gave up. All right, he would go to Wilno.

Early in the morning we went to the railway station. The Physical Society in Warsaw had reserved a car on a slow train which connected the two cities. The distance from Warsaw to Wilno is about two hundred miles; it took us twelve hours to reach Wilno.

Nearly all the members of the society who went from Warsaw to the meeting were professors, lecturers or students at the university. My colleague and I were the only exceptions. I knew that no one would treat us brutally. I knew that all my questions would be answered politely but briefly. I also knew that each member of the group would be glad if I left him alone. I was an intruder to whom they had nothing to say and by whose presence they were not amused. My impulse was to break this attitude, to find some human contact and understanding.

One of the men, known for his gentleness, was polite enough to say a few words to me:

"What will you lecture on in Wilno?"

I grasped this opportunity for conversation:

"It is a paper concerning the connection between Maxwell's theory and relativity theory."

"We experimental physicists are hardly interested in such problems. There is too much theory in physics. The essential thing is to make experiments, to observe and to write down data."

Feeling unsure of myself, I began to talk unnaturally and loudly in my attempt to convince him.

"Don't you see that what you are doing is also mostly theoretical? Before you perform an experiment you have some idea of its purpose. You employ a tremendous lot of theory in performing it. You see some cadmium lines but you write down wave lengths, and you must look for a formula connecting the numbers in a simple law. The difference between my theorizing and yours is only a quantitative one."

I should have liked to press the point further, but the experi-

mental physicist who had honored me with a polite question regretted it immediately and ended the discussion.

"I don't agree with you. These are two quite different things."
And then, suddenly turning to his neighbor on his left:

"Did you bring your tennis racket with you? Perhaps we shall
be able to play."

I was left alone. I knew that while I feared this attitude of indifference and hostility I had provoked it. In a desperate attempt to be interesting, I was unnatural and talked loudly. In
my excitement I didn't hear my own voice properly; the lack
of response increased my tension and resulted in new blunders.

It was not quite true that I felt against me a united front of
the Gentile physicists working in the great institute of experimental physics in Warsaw. The atmosphere inside the institute
was exceptionally snobbish, even for Warsaw. The social differences inside the department played an important role. Everybody knew from what family everyone else came. Everybody
knew, for example, that one of the physicists was the son of a
shoemaker; the department had its own social outcasts. At least
some of these outcasts were more progressive and more suscep-
tible to personal contact, even with a Jew.

I well knew the atmosphere which awaited me during the
journey and at the meeting. I anticipated it with a definite feeling of horror. Then why did I go to Wilno? Not because I had
a paper in which scarcely anyone was interested. I did it because,
still deep in my heart, there was a spark of hope that a miracle
might happen and that the outside door would open and I might
escape. And I wanted to do everything in my power to make
this miracle more probable. Why did I go to Wilno? I know the
answer now. But eleven years ago I did not want to know it;
I suppressed the problem and avoided the embarrassment of
looking for an answer.

The man who had arranged the excursion asked me:

"Have you a ticket?"

"Yes, I bought it myself."

"Good! Because the reduced fare is only for university em-
ployees."

I stood in the corridor until the train moved, talking to my colleague, aware of the hostility of the crowd around us. We decided to look for a compartment. We looked through the glass windows for two empty places and for a word of invitation. Our look was either ostentatiously ignored or answered by a remark that all places were occupied. In one of the compartments we saw two young girls.

"Nobody else here?"

"No, nobody else."

"May we come in?"

"Certainly."

I was puzzled that in this compartment designed for eight people we four sat alone. After exchanging names I found the simple explanation. Here in this train, in this compartment, we formed, against our will, a Jewish island, a ghetto in miniature.

My first feeling was that of relaxation. The tension diminished. Here in this compartment, on my way to Wilno, I met Halina. I have a good memory, but I find it difficult to recall what was said during that day's journey. I remember only fragments, but still, after eleven years, I have a clear picture of Halina. I can still see her face, the dark, dreaming, tired eyes, the smooth black hair and the childish mouth. She was not spectacularly beautiful. She lacked the external excitement, the glow which has the power to capture at first sight. Her attraction, which worked slowly, intensified by time, lay in quite the opposite. She created peace and quietness around her. The calm way in which she talked contrasted vividly with my excitement, my way of talking as though to assure everyone around me, and mostly myself, of my own importance.

I was struck by the solicitude of her friend. She took care of Halina, asking if she were comfortable, if she would like to sleep, if she would like to have the windows open or closed. From time to time someone from the other compartments came in to talk to Halina, inquiring when she had returned and how she felt. I asked her:

"Why does everybody nurse you?"

"I was in Vienna for a year; I was sick."

"I was puzzled by you the moment I saw you. I don't remember having seen you before at the society meetings."

And I added foolishly:

"And I certainly would have remembered if I had seen you!"

Halina smiled mildly.

"I came back a week ago. And I thought it might be a good idea to begin work again by going to a meeting. A kind of transition from an idle to a working state."

Trying to be funny, I was impertinent:

"I don't believe that there was anything wrong with your health. You look all right. I bet you are an only daughter spoiled by your parents!"

Her forbearing smile was the only answer. I went still further:

"Are you quite sure that you were sick?"

"I had to take the doctor's word for it."

Sometime later I came back to our first meeting and asked Halina:

"I behaved so badly, so noisily and with such self-assurance. Were you not offended?"

I was astonished by her answer.

"No, I was not offended. I know this kind of behavior. It comes from unfulfilled ambitions and unhappy life."

I suddenly realized that I was happy in Wilno and that I thought more about Halina than about the meeting. I asked her to come with me to the official dinner of the society. When I fetched her I was astonished to see how beautiful she looked in her evening dress. I had noticed before that her face changed rapidly. At times she appeared full of life, glowing with vitality; then suddenly her face would lose its color and lines appear: the recurring signs of past illness. But that night she looked more beautiful to me than at any time before.

"Do you know, Halina, you are beautiful!"

"Don't get excited—I worked hard to help nature."

I looked at her dress of black net over blue and silver and remarked how lovely it was and in what good taste. She said:

"My mother brought it from Paris. I suspect that this is the chief reason for her travels: to pick up nice things for me."

The dinner, the meeting, everything now seemed friendly. When I entered the dining room with Halina I found that this time I was noticed. The head of the physics department (who had always seemed to me the nicest of the physicists) invited us to have vodka with him before dinner. What an honor! And all because of Halina. The strain diminished; I felt the outside hostility much less, or, even if I felt it, I did not care. My contentment gave me more confidence and greater freedom of expression. More quiet, more detached, I found a warmer response and sometimes even evoked a spark of sympathy or interest.

I can now ask myself, "Was it really fair to assume that everybody around was hostile and would have liked to treat me badly?" The question is meaningless. I don't know the answer and I never will. I only know how the atmosphere affected me. And I found it much more antagonistic when I was unhappy, much less so when I was happy. I projected my disappointments and unfulfilled ambitions into the hostility of the outside world. It is an attitude in which the seeds of a persecution complex are already sown. This is one of the great dangers of being unsuccessful. But I found sufficient evidence later that not all the hostility lay in my imagination.

During one of our evening walks through the narrow, curving streets of Wilno I found myself, like everyone around Halina, taking care of her, wondering whether she were cold, whether she felt tired. I asked her how long she had been in Vienna and what was her illness. She told me how the doctors in Warsaw had been afraid that she had a tumor on her lungs, how they had sent her to Vienna where the examination revealed that it was a much more innocent illness. They had used an X-ray treatment for therapy, an exasperating treatment which healed her but left her exhausted and unable to bear any physical strain.

"Who was with you in Vienna?"

"My father came with me in the most difficult times, when the doctors thought that there might be danger to my life.

Later, when I felt better and was convalescent, my mother was with me."

"That is strange. I should think that the opposite would be the natural order."

"Not with my parents," Halina answered. "My father has a very orderly brain and can control himself. My mother is terribly nervous and always responds to my illness with an illness of her own. I have my theory about it. She met my father when she was twelve. She fell in love with him and married very young. Now she tries to make up for her lack of experience by living my life even more intensely than I do. I am their only child. This makes it fairly difficult for all of us."

I felt warmth and understanding in Halina's every word. I suddenly thought: "Life would be worth living for anyone who took care of Halina."

We came back together to Warsaw. I returned to my teaching. With my mind full of romantic memories of the Wilno meeting I called Halina two days later, and she invited me to her home.

She had two rooms in her parents' apartment. I should have felt overawed by the signs of wealth and good taste everywhere, so different from my home in Cracow, where three children and a grandfather slept in one room. But nevertheless I felt perfectly free and happy, and our conversation flowed as easily there as in the streets of Wilno. For the first time, through Halina, I came in contact with a new environment.

I had all the characteristics of behavior of the first Jewish generation which tries to fight its way outside the ghetto, the characteristics which are so difficult to conceal. My childhood friends, my colleagues in the school and in the Union, all belonged to the same struggling first generation. We all saw in our studies a pathway leading out from the ghetto. Nearly all of us were disappointed, having recognized consciously or subconsciously that after our university studies, often in distant cities, after working hard and achieving degrees, we finally landed near the place from which we started.

Halina was of the third generation. What is the difference

between first and third generations among Jews? I once learned with horror that, accepting the white man's attitude, many Negroes with light skin look down upon Negroes with darker skin. But do not Jews act in the same way? Do not many German Jews look down on Polish Jews? Does not the third generation of Jews, whose fathers absorbed the foreign background often in the most superficial way, look with scorn at the struggles of the first generation? Still feeling inferior to the outside world where they are not yet accepted, they grasp every opportunity to assure themselves of their superiority by deprecating others.

It was because I knew this characteristic attitude of the well-to-do Jewish third generation toward an unimportant first-generation Jew that Halina's environment impressed me so much. It was enough to know Halina, to guess that the atmosphere in which she had been brought up was different. But only later did I learn the progressive, intelligent and understanding background of her home.

My memories of that time are intense and fragmentary: Warsaw parks in the Polish autumn, small cafés, more and more frequent meetings, hours of longing, the first kisses, plans and confessions, until my whole world began to center around Halina, until everything else seemed like a picture seen through the wrong end of a telescope, small and insignificant. I lectured on Archimedes' principle, asked questions, took notes, went to meetings, worked in the Teachers' Union, but everything ceased really to exist. Only Halina and my love were real, but it was the reality of a dreamlike world; everything else was a mechanical, automatic reflex, thoughtless and unimportant.

For Halina it was first love. Her behavior differed from anything I had experienced or read of before. I knew well enough all the moves in the chess game played by women and dictated by their instinct: to rouse desire, to refuse fulfillment, to test their strength in small moves, to begin the fight for future domination.

On the day Halina discovered her love she surrendered completely and unconditionally. She placed her whole future in my hands without knowing or asking what I should do with her life. Her desire to give, to give everything and not to ask for anything, her strong emotion—all were in the most vivid contrast to her physical weakness.

Love recreates the world. The sun, the streets, the dirt, the faces, everything was different. And, above all, my thoughts of the future changed. I found myself believing that something might happen which would alter my life. From the radiation of love a new store of force developed for the upward struggle.

Halina strengthened this belief in me. We talked of her work and mine, of my scientific plans and her intention of getting an M.A., of small details of our future life, of social questions and women's dresses. We mingled great and small problems; nothing was too important and nothing too insignificant for us to discuss. As a matter of course marriage was assumed in our excited talks.

Once Halina said, using, as she often did, the nickname she had given me:

"Vik, we have decided everything for ourselves and you have not even met my parents; they do not know anything and are puzzled by my strange behavior. I know, Vik, that you won't like it, but you will have to meet my father."

"I hate the idea. It is our problem and it is our life and our responsibility. Why do you accept the bourgeois relationship of sons and fathers-in-law? It is ridiculous. What do I care for your father? He may be liberal on the surface and may give money for political prisoners, but he is rich. He knows the power of money and will look down on me."

But she insisted:

"Don't oversimplify things. My father has a good brain, and I am sure that you will like each other if you don't set your mind too strongly against him. It is true that he makes money, but he has a radical past and is still proud of it. And I just cannot create a *fait accompli*. You will have to go through it. I will make it as painless as possible, and I will tell him our story before he meets you."

Grumbling about the unnecessary and bourgeois invention of parents, I agreed to meet them. Both of them came one evening to Halina's room, and I was properly introduced to an older gentleman with a very pleasant and intelligent face and to Halina's mother who looked young, shy and nervous.

I was struck by the resemblance between Halina and her father. The same long face, the same peaceful expression, except that his eyes were gray, not as dark as Halina's, and there was beneath the surface of quietness an undercurrent of worry and fatigue. His small half-gray beard, his progressing baldness and high forehead gave him the look of the Russian revolutionaries of another generation.

I sat comfortably on the sofa, prepared to enjoy the situation, and decided spitefully not to say a word. I thought:

"Let me see how they will start."

Halina's father broke the silence:

"Halina told me the story, and I know that we all feel very stupid at the moment but we must go through with it."

His nervous way of talking was in vivid contrast to his peaceful appearance. Carefully he chose the most appropriate words and, with broad gestures of his hands, linked them together into rounded sentences.

"Up to now we have been the nearest to Halina's heart. From now on you will take this place, and although my wife and I have often thought about it and wished it, it is a shock when her marriage is actually planned, and perhaps we shall not be able to act and behave as we should like to. But I can assure you of our best intentions to make a success of our mutual relations."

I felt ashamed. I was prepared to find mistrust, at least something which could justify my previous obstinacy. I looked toward Halina's mother. With a handkerchief in her hand she was drying her tears and trying to conceal her emotion. Only Halina, undisturbed, smiled happily.

I was touched by what Halina's father had said, but I did not want to show it.

"I don't see any reason for tragedy. I know from Halina that the most important thing for you is Halina's happiness, and Halina is happy now and I am sure that we shall be happy together."

He answered:

"No. It has nothing to do with tragedy. We are only much more tense about everything concerning Halina than most parents are. Perhaps it will be good to explain why."

And he told me the dramatic story of Halina's illness, of their fight for her life and health. I melted completely and I wanted to go to Halina's father and press his hand. I said:

"I know that Halina needs care. If I were not able to take care of her I would be deceiving both of us now. I believe not only in her future health, but I believe that Halina will regain full strength and will be able to work normally to earn her living as everyone ought to."

I thought that at least here Halina's father would protest. But he only said:

"I am happy that you said it. Halina is tired of hearing from me about the necessity of working and making a living for herself."

And he added, with a smile which broke the tension:

"I hope she will find it more difficult to fuss about your repeating the same things."

I felt warmth and sympathy toward Halina's father; these feelings increased later and cemented a lasting friendship between us. He was one of the finest men I have ever met in my life.

II

My work in the school, all the trivialities of my daily life, now seemed so different against the background of our marriage. We easily made the adjustments demanded by the problems of everyday life. I began to work, at that time on quantum mechanics, and Halina prepared for her last examinations.

And then Halina's illness broke out. It was scarcely three months after we had started life together. It was a strange illness. At the beginning we paid little attention to the symptom, convinced that it was due to the recent excitement and that it would vanish quickly. But it became stronger and more frequent. It was a difficulty in swallowing. Somewhere in her throat the food was stopped and rejected by a contraction of the tube. We were sure that it was only a nervous reaction and that a sedative would help.

We went to a doctor. He took the account seriously, made many X-ray pictures during trying hours, to find out where and when the food was stopped. Then we consulted a neurologist. The X-ray pictures showed that near where the esophagus joins the stomach there was a contraction, preventing the food from entering the stomach. It was for the doctors to say whether this contraction was organic or nervous. To me it was infuriating to see the doctors, not knowing what it was, trying to conceal their ignorance, shift the responsibility from one to another. They knew Halina's parents and the history of her previous illness, and they were afraid to make a definite move. Finally they chose the most convenient way and advised Halina's parents to send her to the famous doctors of Vienna.

Even before the doctors suggested Vienna we suspected that sooner or later they would do so. The thought of Halina remaining for months in Vienna while I stayed in Warsaw and taught school grew as a nightmare between us. After one of the first visits to the doctor Halina, pressing my hand nervously, said as we drove home in a taxicab:

"Vik! I feel that they will try to send me away from you. It will be terrible for me and for you. We must not separate. They do not seem to know what it is, and the most convenient thing for them will be to send me away. My father is so scared that he will agree with the doctor. Vik! Promise me that you won't let me go!"

I was horrified at the idea of our separation.

"Don't worry, Halina; I am sure that your whole illness is purely nervous." I really believed in my statement. "And if so, then we must be able to get rid of it by exercising our wills, by being clever and eating when you don't feel the contraction, by having a peaceful, quiet life."

"You see, Vik. You will have a difficult task with my father. He will have one argument: my weight. The doctors in Vienna said that I must not lose weight. They will insist on checking my weight every few days, and if it goes down they will work on us until I go away."

Because of Halina's previous illness I understood that this was the major problem. Her difficulty in swallowing was not painful, but the result was a continuous loss of weight. There were intervals of relief when the contraction vanished, which convinced me that the illness was nervous, not organic. Each day focused upon these moments of relief. Halina was determined to do everything to conquer her illness. Often as we walked through the streets of Warsaw she would suddenly say:

"I feel better now. Let's go quickly to a restaurant."

At the oddest times of day we rushed to the nearest restaurant, ordering the most easily digested and nourishing foods. It usually began well. Then the nervous strain of quick eating had its effect. I would look at Halina's face and see that the difficulty had started again. A little later she would say:

"Excuse me for a moment."

She would come back relieved but depressed. We would leave the restaurant, followed by the puzzled stare of the waiter. We began to repeat meaningless words to each other. I tried to comfort her:

"Have a little more patience. It will be much better when you have more rest. Your life has changed so rapidly and unexpectedly. You just cannot take it."

"Yes, Vik. It is really nothing. It only looks so terrible. But it is hardly an illness at all. I feel perfectly well; I don't have any pains. I am sure it is nothing serious. The first time, a year ago, when I was ill I was afraid; but not this time."

We decided to check her weight and went to a drugstore. The

indicator showed a decrease: seven pounds less than her normal weight. I tried to console both Halina and myself.

"It is really nothing. The important thing is to keep your weight at this level and not let it go lower. Later we shall think about getting back to normal."

But it got worse. A week later nine pounds less; again, a week later, the total decrease was thirteen pounds. I became frightened. I thought that the present weight must be maintained at any price. The thought of Halina's food became an obsession. I found myself brooding constantly about the number of calories she took each day; every morsel of food was carefully considered and weighed in my thoughts. Food, eating, weight, avoiding Vienna, were the most important topics of our conversation; nothing else mattered.

"Today you had a good day. You drank two glasses of milk and ate a lot of butter. You ate nearly your whole dinner. Only supper was a little worse. If it goes on so, and I am sure it will, you will be gaining weight and everything will be all right!"

But after one good day bad days would come, interrupted by intervals of relief. The scale would again show a decrease in weight. Halina's parents were worried. We tried to discuss politics, literature, external topics, when we were together, but everything was artificial and irrelevant to our real concern. The problem of Halina's weight stood in the center of our lives, and we all knew that it was the only problem that mattered.

It was five months after we had started life together. There was no way out. The doctors in Warsaw did not want to take the responsibility and would only advise Vienna. Halina's father decided to go with her since my school kept me in Warsaw. We all tried to be brave, consoling each other by saying that it would be only a matter of a few weeks before Halina would come back blooming and healthy and all our troubles would be over.

At the station, in the last few minutes of tension, we repeated to each other meaningless and trivial sentences:

"I shall write to you each day."

"So shall I, and there is still the telephone."

"Don't forget to send a wire if there is anything important."

"I am sure they will send you back quickly. Nothing can be really wrong with you. You look splendid today."

"I feel very well. I really don't know why I am going."

The train began to move slowly, and Halina's whole strength was gone with the first turning of the wheels. I still can see her bursting into tears, helplessly trying to send me the last gestures of love and to conceal her emotion. And I felt the emptiness of the lonely days to come and the darkness of the town.

III

A FEW WEEKS LATER a girls' club in the school where I taught arranged a forum to discuss the co-operative movement. According to the rules, each such meeting designed for the students of the school, and for guests from outside, had to be attended by a member of the staff. The girls asked the principal to designate me, and I agreed to supervise the meeting.

The classroom was crowded. I was astonished that a lecture as undramatic as a report on the co-operative movement should be attended by so many guests, most of them boys from other gymnasia. The speaker, an intelligent girl from the senior class of our school, read her brief lecture emphasizing the high development of co-operatives in the Scandinavian countries and in Soviet Russia.

Then the discussion began. Soon I guessed that the performance had been rehearsed. The boys took the floor, shifting from the subject of co-operatives into pure politics, praising Soviet Russia and denouncing Poland as a fascist country.

There I was, a member of the staff in a school supervised by the Ministry of Education, listening to speeches which violently denounced the administration of my country. According to the laws and practice of the courts, each of the boys taking part in the discussion risked four years' imprisonment. Besides that, the fate of the school was at stake.

I saw no sense in this childish demonstration. Above all, I feared that there might be a provocateur in the audience. Poland was full of them. In that case the meeting had become dangerous for the students and for the school. But even if there was no danger of provocation, then for whom was this designed? Outsiders could only be antagonized by the superficial and childish arguments and the arrogant tone of self-assurance with which the worn-out slogans were pronounced. Even intelligent sympathizers were bound to find this disgraceful.

I took the floor and in a moderate speech pointed out that they should adhere to the subject matter, that is, to the co-operative movement. This helped for a while. But once more a speaker began:

"Regarding the co-operative movement, we see that there is only one country in the world where the workers . . ."

I could have spanked these youngsters for staging this aimless demonstration and risking their lives and the fate of one of the very few progressive Jewish gymnasia for nothing.

I took the floor the second time and explained that I would strongly advise them to talk to the point and finally warned that they might force me to close the meeting. Suddenly and without warning I saw myself entangled in a situation which I knew must end badly, regardless of what I might do. The boys ignored my threat and turned emotionally against me. The atmosphere grew more and more tense. They had thought we were all actors on the same stage. But now I had become for them an outsider, a member of the staff, representing the ruling class. I became someone who might, perhaps, be converted. They continued with their immature, excited talk, repeating the same phrases. The actors in this play were young boys, sons of rich Jewish parents who identified Communism with the fight against their environment. Judging from their talk, they

neither knew nor understood anything except the few slogans
which they repeated. Much as I hated to do it, I was forced to
close the meeting. They ran from the lecture room shouting in
the corridors:

"Down with the government!"

"Down with Pilsudski!"

But the moment they left the school they left their radicalism
behind. The street was a dangerous place.

I went home depressed. I was still afraid that it was a deliberate provocation and that the story would come out. On the
other hand, I thought, if I am wrong and, according to my duty,
tell it to the principal, then a few girls who obviously staged
the show will be thrown out of school. I decided to take the
risk and not to tell anyone.

A few days later an article appeared in the most reactionary
and anti-Semitic Polish paper, describing the incident and attacking me for my lenient attitude. I have no proof that the
affair was staged by provocateurs who used the youths, convincing them that they were doing something heroic. But the
fact that an unpublicized meeting in a ghetto school was reported in the news, that my name was mentioned with the
proper spelling, might support this theory.

A government investigation began. The arm of Polish justice
was strong but slow. A year later I was still being called before
the judge to give evidence of what happened during the meeting. The principal, my colleagues, the pupils, feared reprisals
would imperil the future of one of the best Jewish schools. At
least it was clear that the pupils would not suffer as individuals
since it was impossible to collect evidence against them. But the
closing or disqualification of the school would be a purely administrative measure. I felt definitely that this might happen
and that I would lose my job. The outcome would be more
dangerous for me than for my colleagues. My name was associated everywhere with the incident, and the chance of getting
a job in another Jewish gymnasium seemed very slight. A new
danger had appeared in my life, and I had to wait a long time to
discover whether the danger was real or imaginary.

IV

THE MONTHS OF LONGING were coming to an end. Sixteen hours more in the train and I should see my Halina's sweet face. I tried to review the situation once more. I had done this constantly for months in my thoughts and in talks with her parents. From our telephone conversations, from the letters which I had received at least once daily, I could form some picture of Halina's condition. The letters, covered with her irregular writing, were full of the details of everyday life, mixed with words of love and longing. She wrote me about her eating, health, weight and what the doctors said.

The diagnosis of a famous Viennese doctor who had treated her before was that the illness was nervous. He recommended a sanatorium outside Vienna, to which she went. Atropine injections were used to stop the contraction of her throat. The injections removed the symptom for a few hours and made eating possible. But after the atropine ceased to work the spasm reappeared. From Halina's letters I learned that the doctors had succeeded in stopping the decrease of weight by diet and injections. Her weight went up and down in small oscillations, remaining essentially constant. But later the atropine injections, which removed the symptom and not the cause of the illness, had to be increased from the initial half portion to three double portions a day.

On the advice of the doctors in the sanatorium, Halina tried a nerve specialist, with little effect. All this took three months—until the end of the school year, when I had over two months' vacation before me.

Once more, in the train taking me to Halina, I thought for the thousandth time:

"How can Halina regain her health in the atmosphere of longing and suffering heightened by her great sensitivity? Love must be able to conquer bad health. Everything will change now when we start life together again. The essential thing is to finish once and for always with the depressing atmosphere of a sanatorium, the atmosphere of sick people confiding in each other their real or imaginary illnesses, the atmosphere of idleness, where waiting for the doctor's visit or for food and conversation about health are the only pastimes of the long days.

"It must be possible to conquer Halina's illness. We shall go somewhere in the mountains to a comfortable hotel where there is fresh air and good food. Now, being free for nearly three months, I can take good care of her, and she must get better. I am convinced that Halina's difficulties must melt in the high temperature of my love and devotion. There is nothing in her illness which cannot be removed by care, warmth and happiness."

At 7 A.M. my train arrived at the small town where Halina stayed in the sanatorium. Halina was on the platform with flowers, looking helplessly around. We could hardly talk to each other. All the words unsaid during the months of separation, all the suppressed desires, all the weeks of longing and waiting, now stood between us. Only the most banal words came out.

"Did you have a good journey?"

"Are you tired?"

"How are my parents?"

"Did you have breakfast?"

"Why did you come to the station? I did not expect you."

"I could not sleep last night."

We were ashamed to exchange our suffering and love in cheap, trivial words, and I felt that Halina was further from me than on the first day we met. Through all the excitement I realized my dependence on Halina and I felt a moment of resentment. I avoided looking in her eyes. Everything was strange and unreal: the country, the town, the foreign language and Halina against a background to which she did not belong. But the tension vanished as rapidly and unexpectedly as it had come the moment we found each other in her room; after a few hours we exchanged happy, quiet and contented glances, again making plans for our future.

We sat for hours on the terrace of our room, looking out at the big garden of the sanatorium, dissecting our past experiences or trying to penetrate the future.

I had always thought that my childhood was uneventful, dull and colorless and was astonished to hear Halina's remark:

"You had a very unhappy childhood. I thought about it the first day I met you. I even said it the same evening to my friend."

"How did you find it out?"

"I have met men like you before. They belong to the group of, as I call them, 'clever Jews,' who don't find air enough to breathe, who get bitter, conceited and unbearable if they don't get a chance in life."

"I am not bitter now, not since I met you."

"You won't get bitter now. We had luck that we found each other at that blessed Wilno meeting. And imagine, I thought that it would be awful and wondered for days whether or not to go."

During our detailed review of the past I told Halina the story of my military experiences, and when I described my friendship with Samson she asked me:

"Where is he now?"

"By gosh! He may be here in Vienna."

I looked in a telephone book for his father's name; I found it. I called his home. I asked for information about Samson. The answer was:

"Here speaking. Who is it?"

I said my name, but there was no reaction.

"Don't you remember," I shouted, "war, Austria, Franz Joseph, fifth company?"

"Oh! Where are you?"

I told him how to reach me and heard:

"I am coming right away."

Two hours later Samson arrived. He looked as boyish and

thin as ten years before, when I had seen him last. Only the back of his head showed a bald spot shining through the untidy hair.

We tried to cover rapidly the story of the intervening years. Samson was much more interested in the details of my scientific work than in the trivial fact of my marriage.

He seemed to think his story was hardly worth telling. After he returned to Vienna he had studied philosophy at the university, had got sick of it and gone to Palestine.

"What did you do in Palestine?"

"I cut stones on the road. It was terribly hot, the sun shining straight on your head. You are thirsty the whole time. You drink water, warm water. The pipes in Tel Aviv are near the ground and the water is always warm. Then my father asked me whether I would like to come back and finish my studies. So I came here three months ago."

"With whom do you work?"

"With Professor Schlick. He is the best modern philosopher. He is very stimulating."

Samson's way of talking was quieter, more restrained than ten years before. He seemed reconciled to life. Neither of us mentioned our experiences in the Austrian army. It was as though we were both ashamed of our past.

I wanted to learn more about Palestine.

"Why did you go to Palestine? Do you approve of exterminating the Arabs and creating a new Jewish capitalistic center?"

Samson's eyes showed the old spark of sudden anger.

"I am not a Zionist. But how can you talk this way? The Jews raised the Arabs' standard of living. Why don't you go and see for yourself?" And he added less belligerently:

"The Jewish settlement is more democratic than any other in the world. The purchased land belongs to the whole nation. I cut stones, but no one looked down upon me because I was only a worker. And when I finish my studies here I shall go back to Palestine."

I thought it best to avoid discussion at this point. I asked him abruptly about the subject of his thesis. Once we were on this theme we did not leave it. Samson was much interested in the problem of determinism and indeterminism and how the latest developments in quantum mechanics would affect our philosophical views. It was not the old Samson who claimed that there was nothing which could really interest him. He asked me questions about the physical aspect of this problem, jumping into philosophical generalizations and into the question of free will.

I interrupted the long-drawn-out discussion to look at Halina, who lay on the sofa and took little part in our talk. The gray of her face and the dark rings beneath her eyes showed physical exhaustion.

I had to remind Samson that the last local train to Vienna would leave in fifteen minutes. We went to the station together, still continuing our discussion about determinism and indeterminism, when Samson, embarrassed, interrupted me:

"I only had money for a one-way ticket. Can you buy me a ticket back?"

It was the same old Samson. I was more touched by this remark than by anything he had said.

Next day Halina and I went to the famous Viennese doctor who was supervising her cure. Halina wanted to escape from the sanatorium atmosphere as much as I did. I was astonished to find that the doctor immediately agreed with us and recommended that we try a normal, quiet life.

We went to the Semmering Mountains and found a comfortable place where, I believed, we could live peacefully and normally. We sat in the sun, read books, worked on physics and talked.

"Now you will get better. You already look much healthier. Don't try the scale now. It is silly to check your weight every few days. It only makes you more nervous. We ought to wait at least for two weeks. We have a lot of time before us. And then you will go back to Warsaw in blooming health."

Halina began to show the first signs of annoyance.

"Why do all the illnesses cling to me? I know that it is nothing very serious. It is more painful for you to see than for me to go through. It looks so terrible, but really it is not. My previous

illness was quite different. That was serious. I shall never forget
the first day when we came to Vienna. My father took me to
the restaurant, cut my food for me, trying not to show the
tears rolling down his cheeks. The illness now is nothing, but
it messes up our lives. I just cannot get out of the mess. One thing
comes after another, and sometimes I think that there will never
be an end."

"There must be an end and there will be, you will see." I
tried to convince her. "But you must believe in it. The worst
thing is that you don't. I want you to be healthy, to take your
degree, to teach or to do something else in life. I am sure that
I will bring you to the stage where you will be able to work,
but you must help me by believing with me that we must and
shall reach this stage."

Halina was disturbed by my talk. Taking off her dark sunglasses, she played with them nervously and said:

"Vik, you talk exactly like my father. Since I was fifteen I
have always heard that a woman must have her own place in
the world and must work. Only when I was ill did he give up
this idea, I suspect only temporarily. But I know that I can't.
Can you really imagine me getting up at seven-thirty in the
morning, going to a stuffy classroom and teaching for five
hours?"

"No," I answered, "I cannot imagine it now. But we must
have patience and wait. Besides, there are many other things
which you could do more easily. You are so far with your
studies that the only thing to do is to finish them, to take your
final examinations, and then we shall decide what to do. We
cannot build our life on your father's money. You know quite
well how much I dislike the idea. I earn enough for a normal,
modest life. Here we live on a scale far beyond my means. For
the moment I have an excuse. I can say to myself that this is
necessary for the sake of your health. But I really hate it. I
wish we could organize our life according to our means."

"No, Vik; I don't believe that I shall be able to do anything
for myself in life. And I know that this is very bad. I have seen
around me grasping wives and mothers who live the lives of

their husbands and children and become unbearable. They always interfere, annoying everyone around them. My mother seems very timid, but she wants to live my life. I see the danger of not having one's own life, but I cannot do anything about it. I sometimes think that the whole study of physics was a great mistake. I am not good enough to do anything important, and when I begin preparing for an examination an illness comes. Even if not something serious, then it is at least an ordinary cold, enough to knock me out for a few weeks. I believe it would have been better if I had studied languages. This is the only study which I could combine with my bad health. Here I have already learned German. Then I could go to Italy to cure another illness and learn Italian. I could simultaneously collect illnesses and languages and in between time try some translation."

Her last remark relieved the tension.

She smiled at me and put on her sunglasses.

Remembering our talks, I now see clearly that I was trying to bully Halina into health. It was not impatience. I was careful, devoted, and I tried to do my best. But I was possessed by the naïve idea that keeping constantly before Halina's eyes the picture of her future health would make her believe that health would come. It did not occur to me that there might be an opposite reaction, that the picture might be frightening and unattractive, that my vital force might make Halina weaker instead of giving her new strength.

We went for short walks. We were afraid that longer walks might lower Halina's weight. At one o'clock dinner was brought to our room. Halina, since her illness, never ate in the common dining room. We had to be home half an hour earlier for the atropine injection, as it took some time for it to take effect. I watched with fear while Halina made the injection, afraid that the needle might break. Her leg was covered with small scars from the many injections which she had taken. Then dinner was brought in and I waited anxiously for a sign that the atropine had begun to work, afraid that the meal might first grow cold and tasteless.

"I feel some relaxation. I shall try to eat now."

We rushed into our room in the hope that this time the meal would pass smoothly. But nearly always, after a few mouthfuls, the spasm came again and all efforts to finish the meal were fruitless. Often the dinner was hardly touched. Sometimes during our walks, for no reason at all, without the injection, Halina would feel a sudden relaxation. Once during such an interval we rushed to the nearest restaurant and ordered a glass of milk.

"It went down," she said and smiled as happily as if a miracle had happened.

"Try another one," I suggested, and the other glass went down too. We went home in an ecstasy of happiness because Halina had swallowed two glasses of milk.

Thousands of times I repeated:

"Doesn't it prove that it is nothing serious; it can only be nervous? Otherwise how could the difficulty vanish so suddenly?"

"I have never worried about this. I knew that it was nervous. But why is it so persistent?"

Each time something happened indicating an improvement I believed with an unshakable optimism that at last the turning point was reached. Then the symptoms would reappear, turning into despair the hopes that had been awakened.

The trend indicated by Halina's weight was alarming. She had arrived at Semmering fifteen pounds below her normal weight. Our first check showed the loss of seventeen, then of twenty, then of twenty-six pounds. All her clothes had to be changed and changed again. She looked brown and healthy; but the loss of weight, which no external force and care seemed to stop, began to be terrifying.

I would look at the passing women during our walks and try to judge the weight of each of them.

"Did you see the girl who just passed? She is certainly ten pounds thinner than you are and she still looks normal!"

"Yes, she is thinner, she is as thin as I shall be in three weeks. I don't mind how thin I am if we could only stop the decline. You will see, Vik, that sooner or later an operation will be necessary. Why does it all have to be just now, when we could be so happy?"

The problem of weight obsessed us. Food and weight, weight and food, calories, milk, vitamins, fats, dishes, atropine injections, needles and again food and weight danced in our heads. I went to sleep thinking of the food Halina had eaten that day and awakened to wonder: "How will she eat today?"

My optimism began to break. It had been my idea to leave the sanatorium and come to Semmering. I had been cocksure of myself, sure that I was the only one who could help Halina. And now the numbers indicating her weight formed a steadily decreasing sequence. Its law seemed determined and immutable. We tried to joke about it. We talked of how Halina's weight continuously passed all real numbers in one direction. But in the back of our minds there was the terrifying thought of the force which might break the law: death.

We had had almost three months of my vacation before us, and I had been sure that Halina would be cured in this time. Now half of the time was gone and Halina's health showed no improvement. "Perhaps," I thought, "the reason lies in the fear that six weeks from now we shall have to leave Austria and go back to work. Halina is afraid that each day brings her nearer to the beginning of the school year, to the inevitable end of our vacation, to the end of the time during which she must regain her health. If not, she will again have to go to a sanatorium without me."

In Halina there was no bitterness. There was never any lack of interest in my plans; very often I started discussions of my personal problems, repeating the same pattern many times, just to avoid the subject of food and weight.

"For eight years I have taught in gymnasia, six of them, without any possibility of scientific work. I never went through a good school of theoretical physics where I could learn scientific technique; I have never worked in a stimulating atmosphere and I have spent the best years of my life under conditions in which I could only forget what I learned. I feel the results;

sometimes I don't believe that I shall ever amount to anything in scientific work."

Halina understood this difficulty.

"You make one mistake, Vik. You think that you are too old. You are thirty-one. I know physicists in our department who began at this age. We ought to go somewhere abroad for a year, where there are good physicists, and forget your stupid pride in not taking money from my father. He would be only too happy to help. And then when you come back, with some results, you will be able to get a position at the university. They will have to take you even if you are a Jew."

The problem of taking money from Halina's father to invest in my future arose frequently, and I repeated once more:

"I know that with my income I cannot keep you on the level forced upon us by your health. But I at least know that I earn enough for myself. I should feel very badly if I were to give up my job for a year and take a private fellowship from your father. I would take it any moment from Rockefeller or Carnegie, but not from your father. I just cannot do it."

In the night our talk started a new train of thought. Money, gymnasium, studies, six weeks more at Semmering, Halina's health. All the concepts created new associations. Suddenly I knew how to act. Why had I not seen it before?

The next morning I said to Halina:

"I slept badly the whole night because of our problems. I have not been honest with myself. I did not admit clearly that we are disappointed, that your illness is more obstinate and serious than we thought. We believed that by this time your difficulties would begin to disappear, and we must face it: they have not diminished very much. Your weight still decreases and, at this moment, you could not go back with me to Warsaw. We are both terribly scared of the same thing: the end of my vacation. It is a threat to you. Either you will get better or you will stay here alone. How can you get better under such a threat? I thought about it last night. And once I really had the courage to admit all that, the rest was very easy. There is only

one way out. I shall take a leave of absence from my school
for the next year and we shall have a whole year before us. We
can go to Italy for the winter and escape from the danger of
colds. I am sure that after such a year you will be perfectly
healthy. It was stupid pride which prevented me from seeing
this one and only way out: to free you of the threat. I must
not mind taking money from your father as long as your
health is at stake."

Halina answered simply:

"I have been waiting for you to say this. And I am happy that
you decided it for yourself."

I wrote a letter to my principal asking him for a year's leave
of absence, stating that I needed it for "personal reasons." He
approved and was, in fact, glad about my vanishing for a year
since the story of the Communist demonstration, linked with my
name, had begun to be widely known.

We began to make plans for the coming year, and the nightmare of weight and food was, at least temporarily, lifted.

There was certainly nothing noble in the decision to leave my
job for a year. I was bored by teaching in high school, disgusted
by the conditions in Jewish private gymnasia. Here I had the
opportunity of spending a leisurely year in external circumstances which would have been a dream of luxury and adventure
for each of my colleagues. To go outside Poland was a great experience. The rigid and difficult passport regulations, differences
in customs, language and official uniforms, made each country
outside Poland seem exotic. But to be able to go away for a year,
to live in luxurious hotels, was beyond the imagination of those
teachers exactly as it would have been beyond mine a year before. For me, however, it was a decision which I made only after
a great struggle, and I am sure now I was influenced only by my
concern for Halina.

Nevertheless I had an uneasy conscience. In my school, where
spiteful gossip flourished, I could easily anticipate the remarks
of my colleagues by imagining my own reactions to the outward
facts of the situation. I could hear them say:

"Oh, Infeld; yes. He put us on a nice spot here. We have to

clean up the mess which he made at the students' meeting and suffer God only knows what from the authorities. He does not even find it necessary to be here. It is nicer to go to Italy on his wife's money. How people change! And he was always full of radical catchwords. But when it comes to money, radicals love it more than anyone else."

Once as I watched Halina injecting atropine into her leg I saw a sudden expression of fear on her face. The end of the needle had broken off. My strength left me. I became desperate. New complications! New loss of weight! And all because of my stupid determination to do without medical care. I asked the maid to stay with Halina a few minutes while she lay exhausted from the nervous strain and I ran out, distraught, to telephone for a doctor. Before I could lift up the receiver the maid came running. She had found the end of the needle on the floor. Halina and I both became hysterical. We cried and laughed and nearly kissed the maid, who was equally excited by the good deed and the tip she received. But we both felt clearly that this was the end of our experiment. Halina must return to a sanatorium.

Halina entered the sanatorium in Semmering, one of the best in Europe, and I stayed there with her. I hated the atmosphere. For the first time I came in contact with great wealth, exquisitely smart women, men who dressed for dinner; with people half of whom had titles and imaginary illnesses due to dissipation and boredom. The atmosphere was not entirely repulsive; there was some attraction too. I was fascinated by the look of well-groomed women with the sexual challenge in their eyes increased by idle life, mountain air and lack of opportunities. I listened to their excited talks, often in Hungarian or English, of which I did not understand a word, sensing mystery behind the incomprehensible sentences.

I felt that I didn't belong to this world. I was shabbily dressed, since I refused to buy anything new for myself. From the leisurely life, without walks or other exercise, I took on weight. With my small personal income I was compelled to live the life of a very wealthy man, eat rich foods in a dining room where I was by far the worst-dressed guest. I felt like a prisoner in a perverse prison, forced to live sky-high above the level in which he had grown up. I had hated the atmosphere of luxury and soft life before I had ever known it. My contact with this world only deepened my hate. It was a strange world and will always remain strange to me: a world in which material possession counts above intellectual achievement.

Several times afterward I had to repeat the experience, to live among the rich and to test my reaction, which was always the same: the desire to run away. My scorn was not superficial. Neither was it due to the way I was treated. It had seemed obvious to me that anyone would guess that I did not belong to this world, this world where living cost twenty dollars a day. But it was not so. Whoever was in this place must by definition be rich. If I were shabby, did not dress for dinner and did not try to save money on tips, then I must be very, very rich.

Once someone on the sanatorium staff began a very technical discussion with me about the comparative merits of Rolls Royces and Cadillacs. At that time even the names were unknown to me. A car in Poland was a symbol of very great luxury and only a few could afford it. The attendant remarked to me:

"It would have been much more convenient for you, sir, to bring your car with you."

It took me some time to convince him that I had no car.

"Why don't you buy one?"

"I have no money."

He regarded it as a very good joke, laughed wholeheartedly and I thought the best thing was to laugh with him and skip the explanation.

Despite the depressing atmosphere I was relieved that the responsibility for Halina's health was not mine alone. For her the environment of the sanatorium was less trying. She had been there during her first illness; the doctors and nurses remembered her and took special care of her.

The chief doctor, after a careful examination, decided to send Halina to a surgeon in Vienna. He explained why. I did not fully understand his explanation, but the general idea was to insert a rubber stick into the esophagus. Whatever it was, I saw new hope

in it. Budgeting Halina's weight as carefully as one plans expend-
itures, I said, trying to be cool and detached:

"We shall go in a week. We must realize that the week here,
the journey and the first few days in Vienna will take another
six pounds. Then you will weigh a little less than a hundred
pounds. Still not tragic, but high time to do something more
drastic about it."

For the first time since her illness Halina became depressed and
afraid, repeating:

"Let's wait a few weeks more. I won't be able to bear this cure
just now, and I don't believe that it will help me. Imagine a stick
put into my throat. The thought of it makes me sick. If the illness is nervous, then such a mechanical method cannot help, and
if it is organic, then it may help only for a short time and the
trouble will begin again."

Halina, thin, tired, weak and worn out, was at the end of her
resistance. For the first time since our marriage there was a trace
of bitterness in our talks. I blamed Halina for not being courageous enough to go through this trial; she accused me of pushing
her toward something terrible, in which she did not believe.

I insisted until Halina gave in. We returned to Vienna, took a
room in a hospital and visited the recommended physician, a
specialist in throat diseases. Again the complete history of
Halina's illness had to be given, X rays shown, questions answered. By this time I knew everything by heart. We had luck
in finding in our new Viennese professor a man full of charm
and warmth, who knew how to talk to Halina and how to allay
her fears.

"You are quite right. It is awful and painful, especially at the
beginning. The first few times it will make you terribly sick. I
hardly think that I could bear it if I were compelled to go
through it. I can promise you that we shall do it very slowly, and
I shall always interrupt the procedure whenever you are not able
to bear it any longer. And—this is the essential thing—I am convinced that it will help you."

I asked him, when we were alone for a moment, whether there
was any danger. He answered:

"Yes, there is some slight danger, but better not to talk about it. There is nothing else we can do."

When we left the doctor Halina was a little strengthened but still depressed. We went to the medical quarter to buy, as the doctor told us, many rubber sticks of graduated diameters. The largest of them were so thick that I did not believe that they could be used for this purpose. Halina was terrified when she saw them. I called the doctor to ask him whether he had not made a mistake. He assured me that he had not. Halina rebelled again. My words of encouragement, repeated over and over, had lost their effect. She knew them by heart. Sad, depressed, resigned, we went home; Halina went to bed in tears and I with anxiety and a faint hope that the doctor's visit next day might bring some relief.

At eight in the morning the doctor came and I left him alone with Halina, waiting behind closed doors. I tried to force myself to think of other things, but thoughts of the rubber sticks, food, the danger, trooped constantly through my disordered brain. I paced the white corridor of the hospital. Nurses and doctors passed me. Nothing that I saw seemed real.

After a few minutes the doctor let me in. During the first sitting nothing had happened. The path to the locus of contraction would take a few days to penetrate. It was only the beginning, and Halina was too nervous for further experiment. The purpose was simply to accustom her to the unpleasant sensation.

The next day the same procedure was repeated. On the third day the doctor opened the door with a radiant face.

"It went through," he said in an excited voice.

Neither of us knew what effect it would have. Exhausted, Halina lay on the sofa; the doctor told her to rest for half an hour and later to order breakfast. Without saying a word and with an attempt at casualness I watched tensely as Halina tried to eat. For the first time since her illness she ate a hearty breakfast. We dared not comment. It could all have been accidental. But then came dinner, tea, supper. The eating, although far from perfect, was so much better. It was our happiest day since Halina's difficulties had started. Next day again showed an improvement.

For the first time the scale moved upward two pounds. It was the first sign of recovery. The scale, the same dread instrument which I had hated as one hates a spiteful human being, became a wonderful symbol of renewed hope.

With only slight variance the improvement continued. Halina's condition again followed a wave line, but its general tendency was up instead of down. The scale, our good friend, indicated it clearly. New life came to us and we were drunk with happiness, taking taxis all over town, sitting for hours in cafés and restaurants, shopping and spending money, living a gay and happy life.

During one of the following appointments the doctor allowed me to remain in the room. It was dreadful even to watch. The first step was to soak a two-and-a-half-foot rubber rod in hot water for ten minutes to soften it. The next step was to insert it cautiously into the throat, deeper and deeper, until it reached the point of contraction. Sometimes it went smoothly past this point, but more often Halina had to wait, trying to push it again and again, until the contraction lessened for a moment and the rod went through. The end of the curved stick was visible through her mouth. The difficulty was to become accustomed to the sensation of strangulation and to hold the stick inside for a few minutes, then, through practice, to retain it for twenty or thirty minutes. The object of the doctor's prolonged visits was to teach Halina to do this mechanically and easily.

Each day, for the next three years, Halina performed the same operation with the rubber rods. Her difficulty in swallowing, although interfering only slightly with normal life, never vanished completely.

We went back to the sanatorium in Semmering; not for a cure, but for Halina to rest and gain weight. It was autumn. The sanatorium was nearly deserted; the season was over. We enjoyed the comfort and quiet of the place, the last warm autumn days and the relief from tension. We felt happy and talkative again, full of plans for the future and memories of the difficult days just past. We both were sure that Halina was now on the way to a complete recovery.

Budgeting Halina's gain of weight, as we had budgeted its loss, we could safely assume that in two months' time, if nothing interfered, Halina would regain her normal weight. This would bring us somewhere near the beginning of January. Further treatment would be pure waste of time and money. I saw myself caught in a trap. Two months before, in my despair, I had believed it would take a year to cure Halina. But now, after two months, she was essentially recovered. Here was I, with the school year only one month old, without a job, forced to be idle and compelled to live for the rest of the year on Halina's money.

But the situation proved still worse. The long, slow arm of Polish justice finally reached the Jewish school where I had taught. The administrative verdict was disqualification. This meant that the school was no longer the equivalent of a state school, its pupils had not the rights which a state school gave. Hence most of the pupils would leave the school. Soon it must fail financially. It would not be able to provide enough even for modest salaries. Indeed, it turned out that the salaries of the teachers were reduced to only one third their previous amount. For the majority of my colleagues I was the man who had brought about this catastrophe even if not all of them blamed me. How could I return and be compelled to bear the abuse, all the more painful because subtly and indirectly expressed?

I well knew that I was not wanted in my old school and that the school could not provide me with a decent standard of living. For all practical purposes, I had lost my job. And I was sure that the few good Jewish gymnasia were closed to me, not only because there were no openings but also because of the story which had grown out of the students' meeting. I had to face the consequences. And they meant: I was finished as a gymnasium teacher; I was jobless. But I was not frightened. I was rather glad that the routine of my life was broken, that tomorrow was unknown. I believed in myself as never before. Weak and afraid of life, Halina poured courage into me. Her weakness gave me strength.

V

At this time the first insignificant signs of scientific recognition arrived; the world of physics took some notice of my existence. I received letters from abroad concerning the problems on which I worked. There were charming letters from Einstein, encouraging me and illuminating the essential difficulties, in answer to my report on the research I was doing. I received an invitation to collaborate on a Polish encyclopedia.

All this gave me confidence. The pressure of anxiety about Halina's health was lifted. I had a good wife, a companion, and I did not have to go through life alone. By Halina's love and understanding I was raised from the self-consciousness of the first to the much greater assurance of the third Jewish generation.

I found myself hoping for things which before had seemed beyond my reach; I even thought of a position at one of the five Polish universities. It appeared to me that the most logical place was Lwow, a town with the second oldest university in Poland. I decided to work in this direction.

The mathematical department of Lwow University was very good and was known over the world, but there was neither a professor nor a lecturer in theoretical physics. Theoretical physics, a subject required for the examinations, was simply not taught. There was a chair, but it stood empty; there was money, with no one on whom it could be spent. As there was no "Aryan" in the country who could have lectured on theoretical physics, the university decided to take someone from abroad. But again it turned out that the candidates who would accept this job were Jews. The paradoxical situation was that the choice was between no one and a Jew. The choice was obvious: better no theoretical physicist at all than a Jew.

It was a tradition on the Continent that each university should have at least two professorships in physics: one in theoretical and one in experimental physics. The professorship in theoretical physics in Lwow had been unoccupied for several years. There was always the possibility of inviting someone from a smaller university, like Poznan or Wilno, to Lwow. But Lwow was very particular, chiefly because of its fame in mathematics. Since the new professorship was decided by the whole faculty, none of the candidates from other Polish universities could be agreed upon. The vacancy remained, therefore, in Lwow instead of being transplanted to a smaller university. The business of finding someone proceeded with the slowness and dignity characteristic of the Polish universities.

I knew very well that I could not even dream of obtaining a professorship in Lwow. But, I thought, there might be a possibility of a temporary appointment. I had very little hope of getting anything, but I was curious enough to write a letter.

Dr Loria, to whom I wrote, was a professor of experimental physics in Lwow. He differed from the typical Polish university professor who always wore his academic gown, even when in pajamas, and who was full of false dignity and bombastic rhetoric about the purity of science and the greatness of the fatherland. Unlike him, he did not repeat to himself ecstatically every three minutes, "I am a university professor." Here in America, where the relationship between professor and student is extremely simple, it is difficult to imagine the pompous cloud and the air of importance in which the university professor enveloped himself in Poland.

Dr Loria was not provincial in his outlook. For two years he had worked with the greatest experimental physicist of our century, Lord Rutherford, who at that time had been at Manchester. He had spent three years as an exchange professor in the United States and had also done research in Germany. Despite his Jewish blood he had never denied his origin and, most exceptional among Christians with Jewish blood, he not only was not anti-Semitic but had the courage to fight anti-Semitism.

I had known Professor Loria from the old days of my studies

in Cracow where he had been a lecturer. I remembered his friendly, helpful attitude; he was one of the very few lecturers with whom one could talk. Later we met occasionally at society meetings. He had always showed an interest in what I was doing and assured me that he would keep me in mind if there was any opportunity for me outside the gymnasium.

Though I was always thankful for his pleasant manner, I had never taken any of his promises seriously. I was sure that he could not do anything even if he wanted to. But, I thought, I did not risk much by writing a letter. I described my situation briefly, stating that I was free, and asking him whether there were any opportunities at Lwow. At the same time I sent a reprint of a paper of mine which had just appeared.

The reply came by return mail. A friendly four-page letter stating that for the moment he could offer me only a "senior assistantship," but he could promise me his support for later advancement if my scientific work continued. Thus at thirty-one I was offered a job at the university, a position for which I had been hoping for years.

To get a "senior assistantship" is not much. It is really very little. It is a job for a graduate student or for a young man who has just taken a doctor's degree. The limitations of such a job are twofold. First, it is, like all lesser government jobs, very poorly paid. Secondly, it is limited to six years and no one can stay longer; he must either advance to a higher position in these six years or leave. An "assistant" is a man who assists the professor and works under his supervision. The assistant usually conducts seminars for younger students, corrects problems, keeps the department library and often performs some secretarial work for his department.

The funny thing in my case was that I had to assist a professor who did not exist. The entire department, ordinarily consisting of at least one professor and his assistant, was nonexistent. As it turned out to be too difficult to build the department from the top, Loria tried another way, to build it from the bottom, and offered the lower position to me. It was suggested that I assist an empty chair. This really meant that I should perform the work of both professor and assistant for the title and salary of an assistant. I understood this and was glad of the opportunity. I thought that I should have a chance to lecture and conduct seminars, things unusual for an assistant, and still have plenty of time for research. Although I knew that I was too old for this position designed for beginners, I had not the slightest hesitation in accepting it, regarding it as a great opportunity in my life. I believed that I would make good.

It was November before everything was settled. Halina and I were very happy about my appointment. We decided that I should go to Lwow alone, that Halina would join me in a few weeks, in which time she ought to have regained her full weight.

According to the official paper, I was to begin work immediately. When I reached Lwow I went straight to the department to ask Loria what my duties were. We exchanged greetings and decided to meet two days later. I learned my first lesson concerning universities in general and especially Polish ones: Never hurry, one can never be too late. I could as well have come two weeks later and spent the time with Halina instead of plunging into the loneliness of a strange city. Nobody would have raised an eyebrow. I had unconsciously brought the spirit of efficiency and haste from the gymnasium to the university.

Two days later I had my first talk with Professor Loria.

He sat on a swivel chair in his office, playing with a ruler, emphasizing each word, fascinated by the sound of his own voice. The conversation consisted of a monologue delivered with excellent technique and an annoyingly correct diction. I sat timidly on the edge of a chair, listened to Loria becoming intoxicated on the heady wine of his phrases and looked at his thin mobile face and puffy eyes.

"I know that our faculty is much better than others. It is certainly the most liberal faculty of our university and perhaps even the most liberal faculty in our whole country. In this respect you are extremely lucky. But you understand that this does not imply that the faculty likes Jews as professors. Perhaps they could even go so far as, in some special case, to admit an exceptional Jew. What they are really afraid of is that one Jew may try to smuggle in another Jew. Therefore the problem of letting in one Jew is, for them, connected with a terrifying vision of Jews spread all over the faculty." And he finished with a smile:

"They do not know the simple truth that a Jew never lets in another Jew. He always wants to be the only Jew in the place." And then he added:

"If you know the bad qualities of our race and will try to avoid them you have quite a good chance of success. I regard your appointment as only the first step. But everything depends on you, on your work and on your tact."

I was anxious to learn what my work was. I discovered that for the first term it would amount to practically nothing. Nominally I had to conduct with Professor Loria a seminar in theoretical physics which would be, for the time being, the only course on this subject.

So I learned the second truth about universities. They pay very little but they require little work. I had finally achieved what I desired: I had plenty of time for scientific work in the leisurely atmosphere which distinguishes the European universities so clearly from the American colleges. I intended to make good use of the free time and to do my best.

VI

EVERYTHING WAS CHANGED, everything seemed beautiful and full of hope. At the department I had my office, a good library, plenty of free time for research. At home I had warmth, help, sympathy, clever advice. In a few months I went, as I had expected, one step further and was ordered to lecture on theoretical physics. This subject was to be taught again for the first time in five years at Lwow University. I liked Lwow. The town spread over seven hills and valleys and reflected its complicated history in its rich and varied architecture. For centuries this eastern part of Poland had been the scene of wars between west and east. Surrounded by Ukrainian villages, the town had a Polish majority, a Ukrainian minority and a large Jewish ghetto.

The atmosphere in Lwow was still reminiscent of old Austria. No name was pronounced without a title, and a title was always advanced a degree higher than the actual one. Most of my colleagues slept in their apartments, worked at the university and lived in cafés. The correct thing was to go to the same café, to the same table, and to appear there regularly at the same hours, to meet the same people and read the same journals.

Mathematicians had their own table at the Café Roma, where many papers were discussed, written and sent to the scientific journals. Lwow's charm was in its leisurely atmosphere, its superficial quick friendships, its witty and spiteful gossip from which no one was safe and which no one took too seriously. For years the same people met at the same time at the same place in the same cafés, knowing each other's troubles and *affaires*, discussing those of their colleagues but never inviting one another home.

I sit with a friend discussing physics, Bolshevism and the love affairs of our acquaintances, when my companion looks at his watch.

"It is twelve-twenty already. I have a lecture at twelve. I shall come again after the lecture."

Slowly he gets up to pay his bill. The waiter automatically adds the usual tip. Then my colleague shakes hands vigorously with me, we bow deeply and ceremoniously to each other and slowly he goes to his lecture.

Officially lectures began fifteen minutes after the hour and ended on the hour. Only one professor took this arrangement seriously. Known as an eccentric, his punctuality was one of the proofs of his eccentricity. The average lecture began somewhere between twenty-five and thirty minutes after the hour. Trying to be honest and yet not to be regarded as pushing, I began my classes at twenty minutes after the hour.

It was essential for one who was ambitious in his university career to associate with the right people and to sit at the cafés with influential professors. But this had to be achieved slowly and subtly and never by brutal force. It took time and skill.

From looks received and words exchanged one could judge what his colleagues had up their sleeves. Since the university was democratically run, each of them had some influence. Intrigues were usually carefully prepared, and the victim, in spite of the gossipy air, was taken by sudden surprise. As a crude rule I found: if some of the influential men were cordial, it was well to be on one's guard, for they had something nasty up their sleeves.

In the beginning I was bewildered by the atmosphere, and it took me a long time to learn the rules of the game. I was a Jew, occupying a minor position at the university, carefully watched, expected to fail and to go no further. At first I made no effort to meet the right people in cafés. Not because I despised the idea but because I was afraid that whatever I might do would be too obvious. So I met no one, and Loria was the only one among the professors at whose table I sometimes sat. But even he did not invite Halina and me to his home. In a country where university professors are regarded as the cream of society a professor's invitation to his assistant is something unusual. During my first year in Lwow I did not see the inside of a professor's house.

Professor Loria, who watched and knew my work, was—as he often told me—well satisfied with the way I raised the level in the department by my lectures and seminars and with the results I achieved. He began to talk with me about my "habilitation for a docentship." The problem of habilitation was the central problem of my life for the first two academic years in Lwow.

The first step in a scientist's career is to have a Ph.D. This means that, with the help of a professor, one has learned some technique of scientific work by writing at least one paper. With a Ph.D. one could get an assistantship at the university or a teaching position in a gymnasium. The title was, however, by no means sufficient for a steady position at the university. Some years after the Ph.D. the next step in a Polish or German academic career was the habilitation, by which the title of docent was acquired, a title meaning incomparably more than that of a Ph.D. Formally the Ph.D. is open to everyone; the docent title was definitely restricted. The idea was to form a small group of scholars from whom future university professors were chosen. The way to a professorship led nearly always through a docentship. This traditional title meant theoretically that one was allowed to lecture at a university without payment, but practically it opened new possibilities.

Let us take my example. I had an assistantship, a position of little importance. But the moment I became a docent the title would add a new luster to my assistantship. I need not resign from my job after six years. I could keep it for my lifetime and even obtain a small raise because of the acquired title. Thereafter my real title, regarded in Poland as very distinguished, would be that of docent, and the assistantship would become only a means of earning money. I could then live in hope that sometime a professor might die or retire. Then, if I were the oldest among the docents of my subject and not too bad a fellow and not a Jew, I might fill the lowest opening. Each docent thought for years about his chances of going further, recorded the ages and illnesses of all professors in his subject in the five Polish universities, waited patiently, in very modest circumstances, for the good news of an opening through death or retirement.

How to keep Jews and radicals from becoming docents? Every Jew, even before he is born, longs to be a docent in a university. Formally it was a matter of scholarship only. But there was one simple rule which proved effective in keeping out Jews and radicals: before accepting any application, before investigating the scholarly achievements of the candidate, the faculty discussed his personal qualifications. Here was the catch. Here were butchered most of the Jewish and radical candidates. There were many famous stories in Polish academic circles showing how the universities made excellent use of this point.

Once a scholar, too liberal for the taste of the law faculty in Cracow, applied for habilitation. One professor took the trouble to look carefully into the applicant's past. He found that as a graduate student the candidate had written a small pamphlet with a red cover entitled *What Is Socialism?* The professor brought a copy of this pamphlet to the faculty meeting and, raising it above his head, shouted dramatically:

"The man who wishes to be our colleague wrote this a few years ago."

He could not bring himself to pronounce the title. The faculty looked with horror at the red cover and voted a unanimous "no."

The following story is more complicated.

A Jewish medical man, famous all over the world for his scholarship, had a patient who died twenty minutes after Saturday midnight. In the apartment of the widow the doctor met a family friend, a professor on the medical faculty, whom he had known well before. According to the law, the corpse would have to be held in the apartment one day longer because the patient had died on Sunday. The university professor suggested to the attending doctor:

"The death is a terrible strain on the poor woman. The corpse would have stayed here one day less if the patient had died twenty minutes earlier. You could do something humane for the widow if you would write out a certificate that the death occurred a few minutes before instead of after midnight. It would be very nice of you if you could do it."

The doctor agreed and recorded in his official report the time of the death as 11:40 P.M.

A few years later the doctor applied for a habilitation. During the faculty meeting, when the personal qualifications were being discussed, one of the members said:

"By chance I happen to have some information which does not allow me to vote for the candidate. I know that he signed a certificate falsifying the hour of death of his patient against his better knowledge. A man who sacrifices honesty for personal favors cannot be our colleague."

It was the same professor who, several years before, had suggested that the doctor change the hour of death in his report. The faculty rejected the application.

I knew of no case in which an application of a Jew had been rejected explicitly because he was a Jew. There was always the argument: "We do not know his personal qualifications well enough and therefore we cannot admit him to our staff."

Everything was according to law. According to the constitution by which all men are created equal in Poland, there was no official anti-Semitism. Sometimes, however, all the dodges broke down. There were a few liberal fighters at the university, and there were often reasons for which this case or that had to be treated in an exceptional way. Therefore there were some, though very few, Jewish or liberal docents in the Polish universities.

I believe that the following general rule can be stated: if a Jew or radical was admitted to a habilitation, then it happened only when so many "pure" docents existed in the same subject that the Jew or radical would have to wait a century or so for his turn to become a professor.

Later I learned the whole story of my own attempt at habilitation, including much of what happened behind the scenes. It began with Loria, who had discussed this subject with his colleagues and thought that my application would go through. One day he asked me to his office and delivered one of his rounded speeches. He concluded:

"Unless they find that you have raped little girls on the street they will pass you on personal grounds. And all the rest will be easy. My idea is to open the door for somebody who is certainly not better than you but is pure 'Aryan' and to offer him a professorship at Lwow. Then it will be clear to everyone that the chair will not be offered to you. The faculty will quiet down. They will see that I do not intend to push you for a professorship. It will be very difficult for them to refuse you a docentship if they give the professorship to someone who undoubtedly is not better than you. We shall open the door for the other man and then you will slide in too, nearly unnoticed."

Finally Loria succeeded in unearthing a candidate for the professor's chair in theoretical physics. At Warsaw there was a Ph.D., a little younger than I, who had just gone through the formalities of habilitation. Although he was essentially an experimental physicist, he had just written a theoretical paper generalizing some calculations from Schroedinger's to Dirac's equations. This first theoretical paper of Dr "C.'s" later proved to be wrong. This can happen to anyone, but it was unfortunate that it happened to his first paper in theoretical physics when the chair offered him was in this subject.

The formalities of offering Dr C. the orphaned chair went forward. A man had been found with an unmistakably Polish name and ancestry. Later my problem arose.

Carefully I prepared all the papers needed for habilitation. It was a large collection. I had to produce a birth certificate with the ultra-Jewish names of my father and mother, stating my father's occupation as "leather merchant." I could imagine how that would sound to the snobbish faculty members. Then followed a description of my studies and life and reprints of all my papers, a personal declaration stating my religion and nationality, all in all a large collection. I took it to the dean's office, and since the dean was not in I left it with a visiting card. An hour later I received a telephone call:

"This is the dean of the faculty speaking."

"Yes."

"I just got your papers."

There was a silence. I waited patiently. Then:

"I should like to know whether the data which you have given in your application correspond to the truth."

The situation began to be grotesque.

"Yes, they correspond to the truth to my best knowledge."

The voice through the telephone began again:

"You declared in your application that you are Jewish."

"Yes, I am; at least so far as I can recall it."

"That's good; it clears up the atmosphere. There have been rumors that you changed your religion because of the habilitation."

I answered stiffly:

"No, I did not change my religion."

Again the voice through the telephone:

"That's good; it clears up the atmosphere."

To finish the conversation I asked:

"Is there some more information which you require?"

"No, thank you," ended the conversation.

I shall never understand the meaning of that telephone conversation and what was in the dean's mind. There are two possibilities. The first is that in the gossipy town of Lwow rumors were circulated about my leaving the old prophet Moses and chasing Christ only because of the habilitation. The other possibility is that the sentence "It clears up the atmosphere" was prepared by the dean to use when he heard the expected news that I had been baptized. As nothing new came to his mind he used the sentence prepared for the opposite occasion. I know that my two theories contradict each other, but so do many things connected with anti-Semitism and Jews who are at the same time accused of being capitalists and Communists.

The faculty meeting at which my application was considered finally took place; it was near the end of my first academic year. I was sure that the decision for or against me would be taken at this meeting. Next day I saw Loria looking very tired, with dark circles under his eyes.

"I did not sleep all night. It was terrible."

The faculty meetings were secrets, but everyone in town who was interested knew all the smallest details about them. The discussions were too interesting and too exciting for gossip to leave them secret. From many sources I heard the details of this meeting and assembled a clear picture of what happened. When my application was reached on the agenda Loria officially presented the outline of my case. After properly exaggerating to the faculty how charming and capable I was he introduced the motion that my habilitation procedure be opened by approval of my personal qualifications. One of the professors took the floor:

"The problem of this habilitation is very peculiar. Ordinarily the scholarly qualifications are judged by a small committee of professors of the same or closely related subjects. But we do not have a professor of the same subject at this moment. We shall have difficulty in judging the scientific work of the candidate. It is, therefore, illogical to go through with a vote on his personal qualifications before we know how to investigate his scientific qualifications. I suggest that we reverse the usual order and discuss who will judge the candidate's scientific papers before we discuss his person."

All this had been staged in advance, and Loria was taken by surprise. He defended his position:

"According to law, the first step is to take a vote on personal grounds. The second step is to choose a committee which will prepare a report about the candidate's scientific work. I have never claimed that I am competent to judge work in theoretical physics, and it is true that we may have to ask someone outside our university to give us his report. We ought not to wait until Dr C. officially joins our faculty, because it may take some time until this happens. And besides, it would be tactless to leave the matter to Dr C. who is younger and less experienced scientifically than the candidate. In any case the law clearly indicates that this is the *second* step and ought to be discussed only after the personal question is settled."

The issue of discussion was: what to do first. The faculty meeting was the stormiest ever held. Loria was caught in a trap; he saw that everything had been decided behind his back, and he lost his equilibrium. Beginning with logic, the discussion ended in a scene of bitter emotional accusation, with Loria pounding his fists on the table, throwing his chair and banging the door loudly behind him as he left the meeting before the discussion ended.

The faculty decided not to settle the personal issue at the moment but to send my papers to two professors—to my old Professor N. in Cracow and to Professor M. in Warsaw—and to vote after the reports from the two professors were received. It was a clever move, since it was difficult to disqualify me on personal grounds. I had not written a pamphlet about socialism, I did not sign a false death certificate, and they could hardly claim that they knew nothing about me after I had worked a year at the university. The embarrassing incident of the Warsaw student demonstration had not become known in Lwow. So the clever thing for them to do was to postpone the issue. Loria, who regarded the whole battle as lost, explained to me:

"They did not vote on the personal qualifications because they had not made up their minds on which of the two hooks to hang you. If the reports are unfavorable, and they will do their best to make them so, they would much prefer to hang you on the second hook, saying that you are a nice man but a bad physicist. If the reports are very enthusiastic, then there is still the first hook unused: they will claim that you are a passable physicist but a nasty man. I am so bitter about the affair that I should love to resign if I had some financial independence."

Then, as he often did, he turned the conversation to his memories of England and America and how unwise he had been to allow his patriotic feelings to interfere with his career and to return to Poland when he could have stayed in the United States.

For me it was now certain that my career at Lwow was, after all, a failure. I was convinced that the faculty would succeed, by one method or another, in confusing the issue and in preventing my habilitation. Two professors were elected to give opinions of my work. One was my old Professor N. from Cracow, who had refused to do anything for me before. Would he now have the courage to take on his shoulders the responsibility of promoting a Jew in another university when he had clearly declined to do anything for his own pupil at the university where he was lecturing and had influence? I knew the other man, Professor M., very little. He was an experimental physicist with some knowledge of theoretical physics. Neither of the two was interested in or acquainted with the kinds of problems I was working on. They disliked one another, and each of them discredited the opinion of his colleague.

I thought:

"The formalities for the future professorship of Dr C. went through the faculty without any difficulty although he is younger and certainly not better than I am. But for a docentship, which means so much less, the faculty invents a whole machinery for obtaining outside opinion, for disregarding the lawfully established procedure, only in order to kill my application.

If only the professors would show the same ingenuity in their scientific work which they display in fighting such cases as mine, Polish science would bloom and flourish."

I felt again encircled by the hatred which had closed doors before my eyes just at the moment I had caught my first glimpse of the desired world. I wanted to leave Lwow. But Loria persuaded me to disregard my pride, to go through the fight to its end and not to give up.

Loria was not the fighting type. He had great charm, intelligence, understanding. He made friends easily, was honest, loyal, but a little opportunistic. But here in the struggle about my docentship a new man was born: a fighting and determined Loria. He believed that injustice had been done to me, that I was treated badly and that it would be good for the department if I stayed. I don't think he saw that this particular problem was a manifestation of a much more general social problem and that while fighting for me he fought for something more important and general than my future in Lwow.

VII

MY SECOND and, I was sure, my last academic year in Lwow began. Two years had elapsed since the meeting of the Physical Society in Wilno where I had met Halina. Now the next biennial meeting of the society was planned for Poznan, another university town. I decided to act as though nothing had happened, to go to Poznan, lecture on my new paper and bravely meet the hostility of the outside world with the increased self-assurance which I owed chiefly to Halina. In grim determination I went alone for two days to Poznan, avoiding personal contact with physicists, trying to show by my presence that I cared little and by my isolation that I didn't intend to be pushing or insistent.

During the second day of my stay in Poznan I went to a restaurant which looked decent and was far enough from the place of our meetings so that the chance of finding any physicists inside seemed small. I had begun my meal when Professor M. from Warsaw, the same professor who had been asked by the Lwow faculty to write an opinion on my work, entered the restaurant. He was very fat, with a large spherical face and a triple chin. There was a spark of cruelty in his small, half-closed eyes. When he saw me he rushed to my table, shook hands with extreme warmth and decided that we should have our meal together. He ordered a beefsteak.

"Beefsteaks are very good in this restaurant."

He licked his lips. Then we talked of the meeting and the scandalously low level of theoretical physics in Poznan.

He ate the beefsteak with tremendous speed and ordered another one, repeating:

"Beefsteaks are very good in this restaurant."

Then he ordered a big beer, became angelic, satisfied, and began an interesting theme:

"Do you know that the Lwow faculty asked me to write an opinion of your papers?"

"Yes, I know it. You and Professor N. in Cracow."

Professor M. melted more.

"I liked your papers. You have talent for theoretical physics. I am now working only in experimental physics, but some time ago I did something in theory myself."

After each sentence he stuck out his tongue and licked his lips.

"Certainly"—I showed my erudition—"I remember your beautiful and important paper on radiation."

"Yes. Einstein liked it very much, and it is still quoted. So you see, I have intuition. Even if I am out of theory I know more than some theoreticians who are supposed to lecture about theory. And I don't mind repeating: you have talent for theory. The only trouble with you is that you have been abroad too little. I know how much it means. In Leyden, where I was invited to work, there was a fine atmosphere. You ought to go to Ehrenfest in Leyden. He is a difficult and strange man but most helpful if he likes someone, and I have a hunch he will like you."

"It is very charming of you, Professor M., to say all that, and I am sure you are right. I know my limitations and I know that I am a self-taught physicist. But to get a fellowship somewhere one must have a university position. Nobody will give money to a gymnasium teacher or to anyone outside the university."

Professor M., looking sadly at me, began with a deep sigh:

"You see, Infeld, I know you little but I always liked you. I will talk with you quite frankly. You know that I am not anti-Semitic. I have been around too much and some of my best friends are Jews. Only provincial professors, not knowing what Europe really is, sitting down here, growing in local greatness, can be so much against Jews. So I must tell you quite frankly: you will never get a docentship; it does not matter what I write about you."

"Yes, I guess you are right," I admitted. "But just the same it was very nice of you to do your best."

"I must tell you the truth," began Professor M., feeling a little bad and sticking out his tongue more than usual.

"I did not write whether you are or are not qualified to be a docent. After all, the faculty did not ask me about that. I wrote many nice things about you and that you ought to get a fellowship abroad."

I never expected Professor M. to play this trick. It was obvious that what he really wanted was a clear conscience and to avoid committing himself, lest he be responsible for introducing a Jew. He well knew for what purpose his opinion was asked. Instead of giving a decisive "yes" or "no" he ignored the question of my habilitation, stating that some further education would do me a lot of good. I became bitter and forgot that I ought to be humble and honored that a professor ate at the same table with me. I asked him outright:

"You read my papers. What do you think: do I deserve a habilitation or not?"

"You certainly do."

"Then don't you see that by your action you have given the faculty a wonderful excuse for killing my application?"

Professor M. was not offended at all. To my great surprise he was rather apologetic.

"First of all you know that I am not the only one. The other professor who had to write his opinion—whom, as you know, I dislike immensely and who is a coward—certainly wrote in such a way as to kill your application. And there is another thing I must tell you which I just learned from a professor in Lwow. They say that you are a Communist. You know, for them every Jew is a Communist. So they will kill your application anyway."

He became really depressed and sorry for me.

"I wonder whether I could help you. What a pity that I did not think before about writing clearly that you deserve the habilitation even though nobody asked me explicitly about it."

The atmosphere was rather pleasant. Professor M. lied to me in a kindly way and he knew perfectly that I knew he was lying. He proceeded in a friendly fashion:

"You ought to do two things if you want to change the situation. First of all you ought to make a public statement that you are not a Communist. Secondly you ought to ask Loria to ask the faculty to ask me a direct question whether you deserve habilitation or not. And then I certainly will say 'yes.' "

"I am sorry, but I won't do either of the two things. It is up to the faculty to prove that I am a Communist and not up to me to prove that I am not. And as for asking Loria to ask the faculty to ask you, it seems to me senseless; officially I don't know anything about the whole business although all Lwow and Poland is talking about it. And besides, you did what you thought was best, and I should hate to make further use of your kind generosity."

In plain English the last sentence meant:

"You have behaved in a dirty fashion already, and I don't like the idea of having anything further to do with you."

We understood each other perfectly through the flowery language, and our conversation ended at this point.

In a town like Lwow, where gossip travels with lightning speed, where the problem of habilitating a Jew is something which stirs the whole town, gossip can be a two-edged weapon. The weapon used against me was:

"He is a Communist."

I made use of the gossip by announcing right and left:

"I have heard for sure that I will not pass the personal voting. The faculty has decided to disqualify me as a Communist because the professors haven't the courage to state clearly that they don't want a Jew."

In my bitterness I had hit upon a weapon which proved a clever diplomatic trick. A few decent professors in the faculty perceived the hypocrisy, and a nucleus of sympathetic voices was formed. The plan to brand me as a Communist was exposed. Things must be done artistically in Poland. There would be no fun in branding me as a Communist at the faculty meeting if Loria and others could anticipate it. The element of surprise would be lacking and this would spoil the game.

The result was startling. After the opinions of the two professors were received the meeting was held and I was approved by everyone on personal grounds. So far as I know, I was the first Jew who went through the personal rating unanimously. Neither Loria nor I rejoiced in it, however. We understood that the decision was to hang me on the second hook, which meant that I was not good enough scientifically to be a docent of the university in Lwow.

Simultaneously a whispering campaign was started about the reports which came back from the two professors. It was arranged very cleverly. The dean of the faculty called Loria and told him blankly:

"The reports concerning Infeld's habilitation came back. I cannot show them to you, but they are not what you could call favorable, if you understand my meaning. If I were you I should advise Infeld to withdraw his application. It is much better for him if it appears that he resigned and withdrew voluntarily than for everybody to know that he is not good enough."

Loria came to me and reported the conversation with the dean. I said:

"I am beginning to be sick of all this. If you don't mind, I should rather withdraw."

Loria tried to calm me.

"You must not do anything now by yourself. After all, we are both in the same boat. The situation this year is a little better than a year ago. There are a few professors on the faculty who will behave decently on this issue."

He looked at the names.

"You are quite lucky that the two most reactionary anti-Semites retired this year. The new professor of philosophy is, we know, very progressive."

And so we went through the list, finding that if I were defeated, then, at least, it would be an honorable defeat. Again we decided to fight and not to give up. I don't know exactly how and why it happened, but a change began to be evident. My issue was one of many on which the faculty was divided. I sensed a different atmosphere through the cafés. Probably it was the spirit of fight and fun so admired by Poles which made them interested in the problem. At the beginning everyone thought, "A new bother with a new Jew." But as time passed and we demonstrated our nerve and diplomacy the show began to be enjoyable. The faculty split into three parts on this and other issues: a progressive and a reactionary section, with a centrum voting either way. Even the dean became less hostile.

At the next meeting the dean announced that the opinions from the two professors had arrived and that someone must prepare a report on them.

Since the formalities over Dr C.'s professorship in theoretical physics had not yet been completed by the Ministry of Education and the chair was still unoccupied, the only man who could do this was Professor Loria, the only physicist on the faculty. Finally Loria got hold of the two reports sent from different parts of Poland; on their contents and interpretation depended my future in Lwow.

All such reports are similar. To impress the receiving faculty with their erudition both professors wrote brief summaries of my publications. This filled quite a few pages, an easy task since the summaries were lifted from my papers. Then some noncommittal favorable remarks, like "interesting," "original," and noncommittal adverse remarks, like "only one of many attempts," "not quite convincing," were scattered through the reports.

I could have foreseen that the two reports would be a mixture of good and bad adjectives, both kept at a superficial level. I could have foreseen that both, in their watery criticisms, would avoid stating definitely whether or not I deserved habilitation. But there was one thing which I could not have foreseen. It was unexpected and perhaps the decisive factor in the game.

The two professors wrote their opinions independently. Because of their mutual dislike they did not consult each other while writing. The result was amazing. The good adjectives mentioned in the two opinions strengthened each other. In one I was "well educated," in the other "original." Thus, taken together, I was "well educated and original." But the bad adjectives canceled each other completely. The canceling was so perfect that one might have suspected a deliberate joke. For example, to the sentence: "The author considers too much the mathematical part and too little the physical part of the problem," there corresponded a sentence in the other's report: "It would have been better if the author had concerned himself more with the mathematical and less with the physical part of the problem."

Again the sentence: "The problems of his interest are too restricted," was canceled by a sentence in the other report: "It would have been better if the author had tackled the problems in a narrower field."

These negative remarks were a gift from heaven. Loria played his cards masterfully, emphasizing with humor and irony that the remaining picture, after superposition of the reports, rather did me some injustice by being too good. In his less enthusiastic opinion I was only a so-so scientist, a statement followed by his motion to qualify me for the docentship. This time the faculty was taken by surprise. Nothing relieves tension like a touch of humor. The bombshell exploded into confetti. It was difficult now to vote against me. No one did explicitly. When the cards with the votes on them were counted they read: eighteen yes, two blank. Thus the second step of the habilitation procedure was completed.

The formalities were by no means over. It was not enough to pass the personal and scientific tests. The next step was a "colloquium," something like an examination. While a faculty meeting is held inside the dean's office the trembling candidate, dressed in tails, waits in the hall. He waits until the colloquium is reached on the agenda. He may try to determine from the agenda when it will come, but he may make a mistake as I did. One of the points of discussion which I thought would take a few minutes took an hour and a half of violent quarreling, during which I had to wait in my stiff evening shirt, correcting my silly tie and holding tight to my knowledge so that it would not escape.

Finally the attendant announced that I might enter. The faculty looked at the strange specimen. According to the rules, anyone could ask any questions he wished. There was nothing in the laws to prevent the professors from asking me questions about sex or the advantages of good plumbing.

Loria started the procedure. Absolutely no one was interested in his questions or my answers. Everyone knew that this part of the performance would go smoothly. Quite ostentatiously other professors began to talk to each other, walking through the office, changing places, taking very little notice of me and my rounded, smooth sentences concerning the historical development of relativity theory. Loria finished, and then the dean asked to whom he might deliver me. Raising his hand, not waiting for the dean's permission, one of the professors of mathematics jumped up impatiently and moved vigorously toward me.

The moment he sat down near me the air changed. The whole faculty froze in its momentary configuration. Talking suddenly stopped. The air was so heavy with quietness that I could scarcely breathe. Forty eyes were directed toward me. The mathematician shot his first question in my direction.

It is very easy for a mathematician to formulate a question which another mathematician cannot answer. It is the obvious result of specialization. Fortunately I knew the weak points of the inquiring professor. Every student of mathematics at Lwow knew that this professor formulated questions unclearly. I had no doubt that the professor represented the hostile faction of the faculty. Glancing over the twenty faces, I noticed irony and tension, relieved by a few friendly and encouraging smiles. I felt that my behavior was more essential than my answers.

Trying to be calm and unconcerned, I refused to understand a question until it was clearly formulated, until I knew exactly what the professor asked. After a few minutes' talking back and forth we got the question clearly isolated and well formulated, but it so happened that I did not know the answer. I thought the best strategy was to exaggerate my own ignorance and I said:

"I finally understand what you mean, but I haven't the slightest idea how to answer the question."

The same procedure was repeated. Specialized, unclear questions from differential-integral equations and group theory were shot at me. I isolated and cleared up the questions only to admit my ignorance. The public ceased to be amused. The performance would have been much more entertaining if I had tried to talk about things which I did not know well. But I understood that just this was what must be avoided at all costs. Beginning to feel uncomfortable, the professor wanted to show that there was no malice in his attitude and proceeded with questions concerning the fundamentals of statistics and some differential equations in physics. Thereafter everything went beautifully.

After that a crystallographer with a friendly face asked me about the connection between crystallography and physics, and finally a friendly philosopher had an informal discussion with me about the influence of modern physics on philosophy.

The colloquium was finished. Later I learned that the mathematician declared that he was well satisfied with my answers and that he would vote for me. The result was the same as before: two cards blank and the others all said "yes."

Now came the last act. A candidate for docentship always

suggested three titles of lectures from which the faculty chose one. Two days after the colloquium I had to lecture before the faculty. If the lecture was passable, the procedure would be essentially complete. It has happened in the history of universities, though very seldom, that even at this stage the habilitation has been refused.

"The Electronic Waves" was the subject of my lecture. I intended to keep it so general that all the faculty professors would get the principal idea. I believe I know my limitations as a scientist, but I also believe that I am a good lecturer, capable of a clear, simple exposition.

The entire faculty came to the lecture, and the dean allowed the older students to be present. My lecture took about forty minutes, with everyone listening attentively, and after I finished the whole faculty applauded enthusiastically. I felt that my lecture had at last broken the hostile atmosphere. Again a meeting took place immediately after I finished. According to law, the faculty must vote to accept or reject the lecture and then must review once more all the previous stages. Even the two obstinate white cards vanished. Everyone voted "yes." And in reviewing the previous stages it was discovered that, according to the law, only "yes" or "no" votes were admissible, whereas I had received two blank cards. The faculty decided this to mean that I had passed all the stages unanimously. A few weeks later the Ministry of Education approved the decision of the faculty and I officially became a docent of the John Casimir University in Lwow.

Thereafter the attitude toward me changed; the old prejudices were forgotten. Nearly everyone on the faculty behaved as though he had always wanted me to be his colleague. Before, I had been someone who pushed himself in, a Jew whom they did not know and did not wish to know. But once the issue was forced upon them, once they had to meet me at the colloquia, at lectures and at the cafés, they found that I was human after all. It is the famous "some of my best friends are Jews" attitude. Many Gentiles, forced by business or professional relations into contact with Jews, melt and accept them but seldom change their principal attitude. It is more convenient to regard individuals as exceptions than to think independently about the problem and break the general rule built up by tradition, upbringing and thoughtless slogans.

I can well imagine someone accusing a professor in Lwow:

"You are anti-Semitic, you do your best to keep Jews out of the faculty."

And I can imagine the professor's answer:

"This is not true at all. Recently we habilitated a Jew without any difficulty and we did it unanimously, although there were some formal obstacles since the chair was not occupied. We could not have done anything more for a non-Jew than we did for Infeld. The worst thing is that the Jews have a persecution complex and what they really want is to be treated better than non-Jews."

Saying this, he would have believed in his sincerity. It is so easy to have a bad memory when convenient.

Now I met the professors of my faculty freely in cafés and restaurants. From the moment I became a docent I ascended to a higher social level and Halina and I were even invited to the homes of several professors.

The papers throughout Poland carried brief notes about my acquiring the docentship. My old friends in Cracow and Warsaw, everyone interested in education and in university life, knew of it. After the habilitation, when we went to Warsaw for the Christmas holidays to visit Halina's parents, I telephoned one of my old friends who taught in a Jewish school, to exchange gossip and to learn the news of Warsaw. My colleague sounded extremely stiff through the telephone. At first he gave the impression that he was very busy and did not have time to see me, then suddenly changed his mind and declared that he would come immediately for a few minutes' talk. With a grim face he began:

"I am sorry to hurt you, but I must get it out. I regret that our friendship will be broken, but I have a deep prejudice against Jews who become baptized for any career, even for a university career."

I asked my friend:

"How do you know that I was baptized?"

He answered:

"Somebody whom I trust told me that he knows it for sure and that there is not the slightest doubt about it."

I asked again:

"If I tell you that it is not true, will you believe me?"

"Certainly."

Then immediately he switched to the opposite extreme:

"I am terribly sorry that I made a fool of myself. Every Jewish teacher thinks that he ought to be a docent, and if he is not, then it is only because he was not baptized. It is difficult for him to admit that the rule may be broken sometimes. He prefers to invent stories and to keep his good, superior feelings. I envy you that you are out of this foul atmosphere."

This was not an exceptional experience. The Zionist daily paper in Cracow was one of the very few which did not print the news that I had become a docent, being certain that I had disgraced the ghetto by forsaking my religion. The same attitude was so common that sometimes I began to wonder to myself: "Perhaps, after all, I was baptized somewhere under an anesthetic and have forgotten about it." Only much later, when I began work with Einstein, did my old Cracow ghetto begin to take pride in me. The absurd logic was: "Einstein would not work with a baptized Jew; ergo he is not baptized."

I acquired the docentship during my second academic year in Lwow. In the meantime the chair of theoretical physics was occupied by Dr C. and a normal university department restored. My new boss, Dr C., behaved with tact and our relationship was correct. We soon found out that our tastes and interests in physics differed. By tacit agreement we avoided this subject altogether, talking politics and gossiping.

I was aware that, ironically enough, Dr C., rather than I, was the victim of anti-Semitism. He had the bad luck to base his first and, at that time, his only paper on a formula which later proved to be wrong. But hardly anyone in Poland knew about this until the paper was violently attacked in print by a German physicist. Against the poor background of theoretical physics in Poland Dr C. was hailed as a new star, a future Smoluchowski, as the first great young Polish theoretical physicist. He was pushed into a position for which he was not mature enough, great hopes were connected with his name, and the fear that he might not fulfill them was, I am sure, one of the reasons for his later scientific sterility. It was pathetic to see his restless eyes and to sense the tension in him when a colleague, whether deliberately or not, would ask him the most common question among scientists:

"On what are you working now?"

His is a small but typical example of how anti-Semitism may destroy its own purpose. The purpose was clear: to push the non-Jew against the Jew so strongly and clearly that no one could doubt his superiority, to push him into a position for which he was not prepared and, in this way, to promote purely Polish science. But the effect was to destroy little by little that very Polish science, creating tensions and diminishing the possibilities of its development, thus destroying by anti-Semitism the aim for which it was invented. Here on a microcosmic scale, in the scientific career of Dr C., I saw a clear example of a more general rule: hate can only be destructive.

Exciting news came from abroad, especially from Cambridge's Cavendish Laboratory, directed by Lord Rutherford. Great discoveries in nuclear physics were made. A brief note in the famous weekly *Nature* announced the splitting of the lithium atom into two helium atoms when bombarded by bullets of atomic hydrogen. These experiments, performed by Cockcroft and Walton in the Cavendish Laboratory, were the most striking examples of an enforced transformation of one element into another. Lithium can be changed into helium. The most fascinating aspect of this phenomenon is that the energy of the helium atoms in which the lithium split is much greater than that of the hydrogen bullet causing the disintegration.

The next important discovery, made again in the Cavendish Laboratory by Chadwick and announced for the first time in a brief note in *Nature* in 1932, was that of the neutron. Up to then it was assumed that all matter in the universe, everything around us, consists of two elementary particles: light, negatively charged electrons and comparatively heavy, positively charged protons. The forests, our houses, stores, the earth, planets, stars, our bodies, seemed to be combinations of these two kinds of particles. This had been the picture of our reality. But the one note in *Nature* changed this picture. Again, as so often in science, we found ourselves compelled to complicate our original ideas in our attempt to explain the increasing wealth of experimental facts. Chadwick's discovery showed that we are forced to assume at least three kinds of elementary "bricks," of which all matter is built, instead of two. We had to assume the existence of a *neutron*, of an elementary particle approximately as heavy as a proton but not representing an electric charge. The material world, in the eyes of a scientist, became a composition of three elementary particles: electrons, protons, neutrons. It was the most exciting scientific news of the few years past. Neither of the two discoveries had been predicted by theory. It was now the task of theoreticians to find a new place for the neutron in a logically simple and consistent picture of our external world.

I felt the desire to go, at least for a short time, to a center of theoretical physics. The nearest place was Leipzig, Germany, where Heisenberg, one of the creators of quantum mechanics, had founded a very vivid and famous school of theoretical physics.

I easily secured a leave of absence from the university for two months. At that time Halina was preparing for her final examination, after her M.A. thesis had been accepted. We decided that it would be better for me to go alone for a short time—about six weeks—and for Halina to spend this time with her parents. It was our first separation in three years. But Halina's health, still not normal, prevented her from going with me where she would be dependent on landladies and boardinghouses.

VIII

L̲EIPZIG is an ugly, dull provincial town. It was the last year of the Weimar Republic. The air was full of hate and tension. On Sundays I saw brown-shirted parades through the windows of my room; later, the blue shirts of the Social Democrats; still later, Communists with raised fists. Nazi beer parlors and announcements of Nazi meetings bore the sentence "Jews not admitted." Each day the press brought news of clashes and bloody fights between Nazis and Communists.

I did not make many acquaintances outside the university, and at the university I did not reach the stage which permitted talking about anything outside physics. I was shy because it was my first personal contact with really great scientific achievement and because I felt the still-vivid anti-Polish sentiment nurtured by both the government and the Nazis. In the atmosphere of the town I sensed that the opposing forces were equally strong and that the future fight would be cruel and hard and might destroy the whole nation. I imagined that only after a bitter fight would the progressive forces submit. Never did I think that they would give up without a hard struggle or that the catastrophe was so near.

In this sea of hatred and fighting the physics department formed a small peaceful island free of anti-Semitism. Heisenberg's assistant was a Jew. Toward a foreigner from Poland the atmosphere was reserved but correct.

I was impressed by the high standard of the seminar conducted by Heisenberg. It was a new spirit in education, different from anything I had experienced in Poland. The essential thing was to lead the students quickly to independent and original thought. Young boys of twenty-one lectured about original papers, revealing an amazing knowledge of modern physics. However early I came to the department library someone was already there. Whatever I had heard before about hard-working Germans was not a phrase. It was true.

On Tuesday afternoons at three we had our seminars in theoretical physics. They took two intense hours. Here at this seminar I heard Heisenberg lecture on his first and very important ideas in nuclear physics. At that time he formulated his fundamental assumption that the nucleus, the innermost part of the atom, contains elementary particles of only two kinds—neutrons and protons—and that the light electrons are not present in the nucleus. In an excellent two-hour lecture Heisenberg discussed the forces acting between the two types of elementary particles and showed how his new theory fit many experimental facts. At this seminar I met a young professor of mathematics, Van der Waerden, who was also interested in theoretical physics. Both Heisenberg and Van der Waerden, brilliant and famous professors, were still under thirty. This would be impossible in Poland, where a professor is supposed to be past middle age, very respectable, dress formally and regard life as a serious business.

I had with me the manuscript of a paper connected with a problem on which Professor van der Waerden had worked. I talked it over with him. Once during a discussion he said he could prove something which I believed to be wrong. In proceeding with the proof he made the same mistake which I had made when working on the same question, then stopped and asked:

"Is it all right up to now?"

"No. I believe it is wrong." I was sure that in a moment he would find his error. He looked at the blackboard for a little while and remarked:

"It is quite wrong. I am an idiot."

I am sure that there is no professor in all Poland who would make such a remark about himself. I found, too, that not one of the young professors whom I met in Leipzig was conscious of the professor's gown. They played ping-pong or tennis with the students, lectured in summer without jackets and behaved in a way which may seem natural for anyone who has grown up in America but which seemed strange and new to me. However, I am sure that the atmosphere in the physics and mathematics departments was not typical of the German universities. It was due rather to the impact of brilliant young scientists upon this university, creating an atmosphere unique in Germany and similar to that in American colleges.

Two weeks after I had arrived in Leipzig Van der Waerden and I began work together. The problem arose from our discussion of Schroedinger's paper. It concerned the connection between general relativity theory and the quantum theory of an electron. Our collaboration proceeded very well. It was my first real scientific adventure, my first collaboration with a man of great international fame. The realization that this brief visit to Leipzig, these few weeks in a vivid scientific atmosphere, was enough to produce results strengthened my self-confidence.

Since Heisenberg had left for a summer session in America and Van der Waerden had gone to Goettingen on a visit, I decided to shorten my stay in Leipzig, return to Warsaw and plan my summer vacation with Halina. The essentials of the paper had been worked out; Van der Waerden suggested that I prepare the manuscript and that we finish our collaboration by correspondence.

On my way to Warsaw I stopped in Berlin to see my old friend Joseph. Since the days of my studies in Berlin we had kept in touch with each other. His home was in one of the most fashionable districts of the town, in a tall new apartment house praised in the press as the last word in modern architecture. This was a change. I remembered our cheap students' luncheons, how we had used margarine instead of butter and had carefully considered the expense of a subway ride. I looked forward nervously to our meeting. It is a strange experience to leap the gulf of years. We see that someone has become older, fatter, and that his hair has receded. But what disturbs us most is that suddenly we are confronted with the ugly reflection of what passing time has done to us.

When I entered Joseph's apartment and saw him after all the years I was relieved. He was the same old shy, smiling Joseph. But I was astonished at the change in his surroundings. As I learned later, the apartment and its ultramodern furniture had been especially designed by the most famous modern interior decorator in Germany. There was a small terrace with exotic plants and a chess table so beautiful that I would have hesitated to shift figures across its shining surface.

Joseph was married to a very successful doctor. When I met her she was in the final months of pregnancy. They were both full of fears for the future of Germany, where they had taken root and would have to bring up their child.

In this atmosphere of external splendor and inner tension it was difficult to recreate the old intimacy. Joseph did his best, trying to tell freely what had happened to him during the past ten years. He had hesitated for a long time, torn between two desires—to work scientifically or to work for a cause—and had finally decided to work for the Communist party. He had become a party leader and had written a series of books which were very successful. He took them from the bookshelf, showing me the small volumes published under his party pseudonym.

In the past year he had been suddenly kicked downstairs. He was still an official of the party, but his prestige and influence had lessened. It had happened because of a phrase in one of his books in which the party saw some Trotskyite tendencies. This phrase implied (as far as I remember) that Communism in Russia would develop fully only through a revolution in Germany and in the West. He told me this in a detached way, without any trace of bitterness.

"They were right. The sentence really was bad. But by the time the book was printed the party line had changed. Besides, at least ten people in the party read the book carefully before it went to press and not one noticed this half a sentence. But I guess that if one man has to be blamed, then it ought to be I."

We had dinner together. I sat on the streamlined sofa, looking at the magnificent furniture. I wanted to remark that I had never seen an apartment as fine as theirs, but since this seemed utterly trivial I asked tactlessly instead:

"Are you not embarrassed to invite your comrades here?"

Joseph blushed and said that he saw them at the meetings and in the office, but his wife interrupted belligerently:

"Certainly we are not embarrassed. We should like to show them what they could have in a different social order."

Realizing my blunder, I tried to change the subject:

"Don't you sometimes regret that you left scientific work?"

But I saw that I had covered one blunder with another when his wife answered sharply:

"Is it not scientific work that he is doing?"

The political ground seemed to be the safest, so I again changed the subject, ashamed of my clumsiness:

"What do you think? Will the Nazis gain strength in the future?"

There was one touching thing about Joseph. In answer to this and other questions he never tried to sell me the party line. He tried to talk to me as to an old friend, even though what he said was heresy from the point of view of the party. His answer was:

"The party claims that the Nazis have reached the peak of their popularity and that this popularity must inevitably recede. Unfortunately I don't believe that this is true. I tried to modify this line. I suggested we predict that it will go down finally though it may go up temporarily. But I was outvoted. It will make it difficult for the party to explain the trend of future events."

"The Communists," I said, "call Hitler 'the fool with the big mouth.' I read it somewhere. How is it possible that his tremendous party machine was built by sheer stupidity?"

"Hitler is a criminal, a lunatic, but he is not a fool. It's dangerous for our party to call him a fool. Such a description suggests harmless qualities. He is too dangerous to be called a fool."

"You remember"—I plunged into the memories of old times—"how the revolution seemed near ten years ago. You talked about it as about an event which might happen almost any day."

"Yes. It is sadder now. I am afraid Hitler may come to power, of course only with the help and support of the ruling class in Germany and outside."

Even I thought Joseph too pessimistic. I did not believe the Nazi regime was so near.

We sat in silence. Then in a low voice, as though he were afraid of sounding bombastic, Joseph said:

"Of course the struggle will not be over even if the Nazis win temporarily. History cannot be judged through the span of a human life. And we shall win in the end."

IX

IT WAS EARLY JUNE 1932. I returned to Warsaw to meet Halina. After the first excitement of our reunion had passed, all my questions centered, of course, on Halina's health. She lacked the brightness and gaiety which I had expected. Her only complaint was that she felt tired. She was more interested in my state than in hers, returning again and again to the same subject:

"You need a rest. You are much thinner; you look worn and gray. You did not take care of yourself as you promised me you would. We ought to go away soon."

I tried to explain:

"Just now I am in the midst of heavy work. You know how this affects me. I sleep badly and will have no peace until the problem on which I am working is solved. It will be very soon. There are only some details to think out and then I shall have to formulate the whole paper and send it away. Later we shall take a rest. The few weeks in Leipzig did me a lot of good. I am full of ideas for future work."

When we started to plan our vacation Halina complained:

"It was difficult, Vik, to be without you the few weeks. And I feel terribly tired. I talked with my parents, and they are against our journey abroad this summer. They think we ought to go to Zakopane."

I felt that Halina's parents were right.

"We must stop these expensive vacations. A year ago we went to Yugoslavia, two years ago to Italy. It is all far more than we can afford. I don't mind taking your parents' money as long as I know that it is for your health's sake. But it must stop sometime. We must start a normal life. Now, since your eating is so much better and you don't lose weight, we should try it. I have never been in Zakopane and I should like to see it."

"I dislike the idea. You don't know, Vik, what a bad time I had there when I was ill a few years ago. The doctors recommended Zakopane before they could think of something more complicated. Everyone goes to Zakopane. You will find there the same university professors as in Lwow, the same gossip, the same cafés. I know you; you will sit for hours in cafés and it will be difficult to keep you out. It is a busy, snobbish mountain resort, and I am sure that it is not right for me."

I tried to persuade Halina:

"You know that it has the best mountain air in Poland. You may not like it, but it has always helped you. It is certainly not worse than Semmering. Just because you have unpleasant memories you must break them by going there again. You know that it will be quite different when we are together."

Halina ended the conversation:

"Of course I shall go if all of you insist. But you must promise me to avoid social life. I am tired of it."

We remained in Warsaw a week. I worked on my paper with Van der Waerden. Then we went to Zakopane, sometime in the middle of June. We took a room in a big, comfortable hotel. I had a lot of typing to do and was, at that time, too busy for walks or social life. In ten days the manuscript was finished and sent away.

Halina felt well in the beginning. On the streets we met old friends from Cracow. Warsaw, Lwow. The intellectual stratum is very thin in Poland, and one meets practically everyone during the summer in Zakopane. A few professors from the university in Lwow with whom we were friendly arrived there too. Loria's wife was in Zakopane, awaiting her husband who was to join her later. It was difficult to avoid meeting people. Halina rebelled constantly against the social obligations forced upon us, largely by chance encounters in this noisy and crowded summer resort. Once, as we came back late at night from a café, she complained:

"Promise me, Vik, that it will not happen again. I simply don't feel fit enough for this kind of life. You will see how dead tired I shall be tomorrow."

I promised Halina that we would avoid late hours in the future.

When Halina undressed for bed I noticed yellow and violet bruises on her legs. I had noticed some before, and when I asked about them she answered:

"I am careless. I knock my legs and bruise them easily; they vanish quickly and appear quickly again."

But this time it looked more serious. There were many of them, varying in color from pale yellow to dark violet. I began to worry. During the next few days they increased. Some of the old ones vanished, but new ones appeared. We decided to consult a doctor. It was a new illness. According to the doctor, the blood did not coagulate properly and the small arteries were tender. They broke easily, and the failure to coagulate prevented quick healing. Although the disease was not considered dangerous, it required great care. The rarefied atmosphere in Zakopane was not good for this condition. The doctor advised Halina to stay quietly in bed for a few days, until she became better, and then return to Lwow. He believed that if good care were taken of her the trouble would soon clear up.

Halina took the new difficulty very seriously and was depressed.

"I have hardly recovered from my last illness. I still have to use the awful rubber stick every day, and already there is something new waiting for me. Why do all the strangest illnesses follow me? Why did it come just now, when I had hoped that my worst troubles were over?"

Frightened as I was at this fresh danger, I said to Halina, believing every word:

"You have been much sicker before. You know that I did my best to bring you out of your illness and you recovered even in a shorter time than we expected. The air here seems to be bad for you. I promise to take good care of you. Have confidence in me as you had before, and I promise to pull you out of this new illness. It is our last trouble. You must believe that we shall succeed. I am sure we shall."

Halina smiled feebly and said:

"I do believe you."

Halina stayed in bed. I rang up her parents and told them the story. As soon as they heard it they said that they would join us.

Soon new signs of illness began to appear, all of a similar character. Halina wounded her gums while eating; blood flowed in a small, narrow stream for hours. On her arm a small pimple appeared which she carelessly and mechanically scratched; in two days it had increased to an alarming swelling.

On the second of August Halina's parents arrived. They were depressed rather than worried. The doctor assured us again that there was no danger and that only care was needed.

The third of August came. I awakened, dressed and ordered breakfast for Halina. She ate her breakfast and again wounded her gums. The narrow stream of blood appeared on her lips, trickling along her chin into the handkerchief which she held. It appeared again and again. I asked Halina to lie quietly and not to talk. I gave her a piece of paper and a pencil to write down what she would like to have for lunch. She wrote down what to order at the hotel and then below a remark:

"Will this bleeding ever stop?"

I stroked Halina's hand and asked her to lie quietly. Then I went out for half an hour to await the doctor, who usually called at eleven in the morning. I talked with my father-in-law. He was quiet but watchful. Neither of us had any real fears.

The doctor came. Halina's father and I entered the room with him. The doctor unwound the bandage from Halina's arm to see whether the swelling was better. It had grown still more during the night. He looked grave.

"We shall have to make a small incision here. It is nothing, but it ought to be done today."

Halina became nervous. Her whole behavior suddenly became strange and out of character. She said in a loud, excited voice, sitting up in bed:

"No! I don't want it. I don't believe that it is necessary." And then suddenly:

"I can't see. It is dark in the room. I am dying."

None of us took her words literally. The doctor said:

"It is dark because the clouds just passed over."

I added:

"Yes, it suddenly became dark in the room. We all noticed it. You must try to relax and quiet down."

Then the doctor again:

"You will see in a moment that nothing has happened to you. We shall push the bed toward the window and you will have more light."

We pushed the bed slowly toward the window so that the light from the cloudy sky, reflected from the mountains, fell upon Halina's face. She sat up in bed for a moment and repeated more quietly than before:

"I am dying."

Her father came near, pressed her hand nervously and in a trembling voice said:

"Don't you see, Halina, that nothing has happened to you? You were excited, worn out, and you nearly fainted. That is all. You nearly fainted, but in a few minutes you will feel all right. You must be quiet. No one will open the swelling if you don't want it. Please, Halina, quiet yourself. You nearly fainted, don't you see! You nearly fainted!"

The doctor asked me to order milk and cognac. We waited impatiently, as though everything depended on it. Halina repeated:

"I don't want to be cut. It is dark."

Her voice was low and apathetic. The milk and cognac were brought; the doctor mixed them and helped Halina to sit up. She drank obediently and then lay down on her back with half-closed eyes and began to breathe heavily.

The doctor whispered to us:

"She will feel better when she awakes. It was a nervous breakdown. I shall be back in two hours."

We sat down in Halina's room, her father and I, whispering to each other:

"She fainted from the excitement."

"Don't tell her what she said when she wakes up."

"The sleep will do her good."

"She will feel better when she awakes."

"Yes, I am sure the sleep will do her good."

I tried to ignore the strangeness of her sleep but it worried me more and more. She breathed stertorously. I saw that her foot stuck out from the coverlet, the toes bent in a tense, unnatural position. I whispered to Halina's father:

"Perhaps I should call the doctor. I should like him to be here when Halina wakes up."

When the doctor came he looked at Halina and went out, calling me with him.

"There is something serious. Professor L., a specialist in internal diseases from Cracow, is here in Zakopane. Perhaps we can persuade him to come."

He rang up in my presence, and from his answers I guessed that Professor L. refused to come to a patient during his holiday. With shaking hands I took the receiver, stuttering and hearing my strange voice. I cried:

"I am a docent of the university. You must do it for your younger colleague. We are desperate; we need your help."

"I shall be right over."

He came soon. All of us stood in silence, watching him examine Halina. The examination was brief. Halina's father and I went out with the doctor into the corridor.

"Nothing can be done yet. We must wait. A part of her body is paralyzed. If she comes through, there may be some traces left. The chances that she will be normal again are fairly small."

I sat down on the floor. I had only one desire: for Halina to live. At that moment I would have been willing for Halina to be crippled if I could have saved one spark of her life. The fear of her death came suddenly as an intolerable thought. I wanted to believe in God in order that I might hate Him for what He was doing to me. Hatred against the world and sorrow for myself found expression in a loud inhuman cry.

Halina's father took my arms, shook them, shouting violently:

"Pull yourself together. There is my wife in another room; she is tortured by sorrow. I must think about her too."

I felt melodramatically cruel and crazy. I shouted meaningless sentences, mixing them with tears. If I had had the power I would have destroyed the whole world around me. Halina's father shook me nervously, repeating:

"You are a man, aren't you? You must be strong."

I tried to calm myself by thinking: "What can I do to save Halina?" The only thing I could think of was to telephone Mrs Loria, who was herself an M.D. She had always been fond of Halina, and I remembered that she was staying in a near-by hotel. In incoherent words I telephoned her what had happened. I heard her voice:

"God! How is it possible? I shall come immediately and shall try to bring Professor Rose, who just arrived at our hotel." (Professor Rose was a famous brain specialist. It was he to whom Pilsudski later willed his brain in his testament.)

Perhaps the doctors were wrong. I tried to awaken hope by the thought of the new professor coming to help us. I went back to Halina's room. There were her father and the doctor. Halina's breathing was louder and more irregular. I watched her half-closed eyes, her still face and relaxed mouth, waiting for

a miracle of hope. The doctor and Halina's father stood mo-
tionless. The doctor kept his finger on her pulse.

Mrs Loria and Dr Rose came in. Mrs Loria stood silently near
the door. Dr Rose moved slowly toward Halina's bed, took her
left hand, and I heard his first words:

"There is no pulse here."

The other doctor whispered:

"There is still some weak pulse here."

The pulse in Halina's right wrist was the last spark of life
and of hope. Suddenly a rosy color, the color of freshness and
health, moved from her forehead and spread gradually over
Halina's face. And I heard:

"This is the end."

X

SEVEN YEARS have now passed since the day of Halina's
death. For the first time in my life, while writing these pages, I
disclose the details of a day obliterated by years of suppression.

The human mind forms a defensive wall against thoughts
which would destroy its mechanism. In the first days after
Halina's death, in the atmosphere of pain and desolation, my
feelings centered around myself in self-pity.

"Here I am. I am thirty-four. My life is finished. I still have
two months of vacation before me and then I must go back
and lecture. But how can I lecture? No!—I cannot do it and I
won't do it. I don't want to see anyone. Why did I want to
teach at the university? How much better off I would be teaching school in a small provincial town. If it were the middle of
the term, then tragedy or no tragedy, I should have to go back
and teach. Nobody could replace me in the middle of the year.

I would have to teach for five hours a day. I would have to conceal my sorrow, to talk about Newton's laws of motion and Archimedes' principle. A hard life, filled with work. It would have helped. But not now. I do not need to go back to the university. Anyway, what would I do if I did go back? I would have three hours of lecturing a week. What would I do with the rest of my time? Work scientifically? I shall never be able to work scientifically again. I am caught in a trap. I wanted a university career, leisure for scientific work. Now that I have it, what shall I do with time? I care about nothing. I shall never care for anything any more. Time, time, time. Time heals all wounds. Does it? Let's believe that it does. After five years a thick fog will cover my memories. But how to live through those five years? Halina is no more and will never be. Never, never. For God's sake, I must not think about it. Later, later perhaps, but not just now. I must do something. It is four o'clock. At ten o'clock I may go to bed, take a drug and try to sleep. Six hours more. The hands of the clock hardly move. Six hours and then five years. I ought to shave. That will occupy me for a few minutes. And then I ought to dress. Should I wear a black tie? Stupid convention. No. I won't wear a black tie. I wonder what happened to my last manuscript? What do I care? I cannot go back to Lwow. I feel humiliated. I am ashamed. Everyone will look at me with pitying eyes. But why do I think only of myself? My first duty is to Halina's parents. I want to spend my whole life with them. I will do everything to lessen their sorrow. They have nothing, nothing in their lives. No, it is not true. They are still better off than I am. They have each other. But I will always remain with them. I will resign from the university and we will live together, all three of us. The rest of my life I will breathe the memories of my happy past. I will be a good son to Halina's parents. If I were alone, without responsibilities toward Halina's parents, and had plenty of money I could go abroad. To America, to exotic countries, to people who know nothing about me. I could pretend that nothing had happened. I could run away from the memories. New scenery and a fresh start. Yes, either of the two extremes would help: complete freedom or slavery. I am trapped somewhere in the middle. What will happen to me? It will be interesting to watch. An interesting problem. A second ego climbing above my first ego and watching it. Why do I have such a bitter life? Am I really the only one? I was happy. I was happier than anyone. I had my share of happiness in a concentrated form. Therefore it was brief. I must be able to take it. It is easy to say. I can't. I do not want to be unhappy for the rest of my life. But I must do my duty, my duty to Halina's parents. I always believed that nothing good can come from doing things simply because of duty. No. It is not duty. I do want to be with Halina's parents. Our suffering will bind us together for the rest of our lives."

I see the picture of Halina's parents and myself, imprisoned in their Warsaw apartment, crying and grieving, unable to accept reality. I remember how melodramatic my voice sounded. I thought of myself as on a stage, acting a drama half real and half imagined. For days we sat in one room, the three of us, doing nothing. I smoked one cigarette after another, breaking the silence with incoherent cries, torturing myself with memories.

To this day not one of us knows what was the illness of which Halina died. We had endless sorrowful discussions as to whether the outcome was inevitable; whether it was an unavoidable consequence of Halina's previous illnesses or a cruel accident, perhaps a burst of one small vein in the brain, paralyzing her body and causing sudden death. It was an almost academic discussion, with which we occupied ourselves for weeks. It is amazing that my scientific training did not show me then that the discussion was a meaningless game with words.

Halina's father kept repeating:

"It was an accident. In two hours she was gone. My sweet Halina. The loveliest human being who ever lived. It could not have had anything to do with her illness. It was one of the strange deaths which just suddenly happen."

And then more quietly:

"I know the parents of an only daughter. She always seemed perfectly healthy. Once, while playing the piano, she suddenly

died. It was ten years ago. Her parents still wear black, refuse
to see anybody and close themselves forever in their misery."

I defended the opposite point of view:

"No. I don't believe it was an accident. It must have been a
terrible illness which the doctors did not recognize. But once it
started the end could not have been avoided."

We went on and on in these endless discussions, recalling
bittersweet pictures from the past, tearing our wounds. But we
did nothing to solve the problem. We *wanted* Halina's death to
remain an unsolved mystery, we wanted to live in its sad shadow
for the rest of our lives. When I later suggested that we settle
our discussions by writing to the professor in Vienna Halina's
mother protested violently:

"Don't do it. The doctors don't know anything anyway. I
don't want to tell anyone."

Closed in our room, isolated in our misery, we underwent
strange metamorphoses which none of us could have foreseen.

Halina's father made active efforts to return to life. He tried
to transfer all his fears and solicitude toward his wife, to save
her from a nervous breakdown.

Halina's mother, weak and sensitive, behaved for the first two
weeks with unexpected strength and then suddenly broke down.
She fled from her sorrow into a nervous depression, spending
days and sleepless nights in bed, moaning constantly, resenting
us when we tried to return to life and watching carefully to
make sure we suffered enough. Later she broke down completely. The same woman who had been helpful, shy and quiet
began to torture herself and everyone around her. To me she
repeated hundreds of times:

"You will run away from us. You will not be able to stand it.
Promise me that you will not leave us."

I had to assure her so many times that she and her husband
would always be nearest my heart that finally the words lost
their real meaning in the mechanical repetition.

When, after three weeks, Halina's father decided to go back
to his office she held it against him. How *could* he concern himself with work and money? She expected us to stay together

night and day and to suffer together. All her old warmth and tenderness toward me changed to anger and disapproval.

I had six months before me. Two months of my vacation and four months of a leave of absense which I had taken from the university. I wanted to spend at least this much time with my parents-in-law. But days without work are endless. The utter idleness was torture. I spent days in the same room trying to pass the time and not knowing how.

The channel of thought which moved in Halina's mother's mind was so narrowed by the illness that it finally found expression in but a few sentences which I had to hear again and again until I began to fear I should go crazy.

"Do you think I shall ever be well again?"

"Yes, I am sure you will," I would repeat mechanically.

"Do I torture you?"

"Don't say such things. I am glad to be of some help to you."

"No. I shall never be well. It is all useless."

My automatic answer was:

"You must have courage."

Then something a little different:

"You are young and you can work, but our life is finished."

Again I answered, at first with conviction and later mechanically:

"I may be able to work and to go on, but I shall never be happy."

Then again:

"Do you think I shall ever be well?"

"I am quite sure you will."

"No, you don't believe it. You know that I will end my days in a lunatic asylum. Tell me the truth. Do I torture you?"

The heroic atmosphere of standing together, of helping each other, disintegrated bit by bit, destroyed by our unrelieved efforts to maintain it. It was maintained most stubbornly by Halina's father. He tried to strengthen his ties with the outside world, at the same time worrying about his wife and behaving toward me with friendliness and understanding. Once I told him how depressing were the talks, the questions and answers, repeated over and over again for days during the long hours I was left with Halina's mother. He said:

"I know how difficult and trying is my wife's behavior and I do appreciate all that you have done for us. But you must try not to feel any malice toward her. It is her means of self-defense. Her life is ruined and she has escaped into illness. The doctor assures me that it will pass. We must be lenient and understanding. I know how hard it is for you to spend whole days with her, but I hope that she will soon be better."

I had to admit:

"Sometimes I can hardly bear it. The repetition of the same phrases, the possessiveness developed since her illness, get on my nerves intolerably, and it is becoming worse each day. I am ashamed of myself, but I must tell you the truth. It is difficult even now to control my feelings."

He tried to defend her:

"Think how she was before. It was from her that Halina got her warmth, from her that Halina inherited the ability to sacrifice and to give everything to those she loved. Even now she has chosen—perhaps by a strange intuition—the best way to help me. I am forced to worry about her. And this prevents me from thinking of my own tragedy."

I tried to introduce my personal problems.

"After New Year's I shall have to go back to Lwow and start working. I must choose a life worthy of all Halina did for me."

I felt a little uncomfortable using this heroic language, but it was in the spirit of our common experience and of the mood which still prevailed. He said:

"Yes. I understand perfectly. You must get out of this atmosphere and return to your work. We are in different positions, and it would be unfair to close our eyes to it. My wife seems to act as though the one thing she desires is to see you go to pieces and spend the rest of your life in mourning. But this is only because of her abnormal condition. As soon as she gets better she will want to see you living peacefully and working,

going forward, if there is any hope of it in the face of this terrible tragedy."

The first time we decided to go out of the apartment together
I fell down the stairs. Walking through the streets, talking to
other people, was a new and strange experience which I had
to learn. Cafés, parks, streets, taxis, human faces—everything
recalled events, meetings and words, looks and smiles exchanged
with Halina, creating sharp, sudden associations, increasing the
opaque and dreamy unreality of my world.

Slowly I began to feel the challenging stream of life. It began
purely as a device of self-defense: brief intervals of relief, sudden feelings of relaxation and mysterious hope invigorated my
mind only to relinquish it again to melancholy. The desire to
escape from this depressing mood grew stronger. I found that
drinking brought a temporary relief, minimizing the importance
of external events and strengthening the feeling of life: "I am
alive because I am drunk." I went for short walks to take this
medicine. But each time I had to obtain permission from Halina's
mother.

"I must go for a half-hour walk. I have a splitting headache."

She would reply:

"Must you go and leave me alone? Take a headache powder!"

I would protest:

"I took a powder and it did not help. I must go out. I shall
be back in half an hour."

The argument would continue:

"You say half an hour. But you will stay longer. Promise
me that it will be only half an hour."

I promised solemnly that it would be only half an hour. I
knew that near the end of the half hour she would grow restless and constantly watch the clock.

My need to escape from this corrosive wretchedness became
imperative. I talked it over with Halina's father and confessed
that I could stand it no longer. The only way out was to hire
a nurse who would take care of Halina's mother. She protested
violently:

"No! I won't have it! Your next step will be to put me in a lunatic asylum."

Against her protest we took a nurse, a sweet young girl, and Halina's mother began to cling to her emotionally as much as she had clung to me before. I was free, and now I could try to go back to work.

Help came from outside. The manuscript which I had sent to Professor van der Waerden came back with his changes. I had to study them and make my suggestions. Everything had to be settled by correspondence, and this took considerable effort and occupied my time. Also I had some new ideas on the generalization of our theory and started to make calculations. For three days I worked as I had in the old times. But after the third day the desire to work vanished as completely and as suddenly as it had come, and for a whole year, with the exception of those three days, I did no scientific work. I could do research six weeks after Halina's death, but six months later I could not work. The mechanism of my brain became affected slowly. The diffusion of apathy took time until it influenced my whole organism.

I tried to read light books, but even on them I could not concentrate. The idleness of endless days poisoned the hours of my life. I wanted to do something which would not seem as aimless as every motion of my body and every thought of my mind. It could only be something which I should be able to connect with the memory of Halina. Suddenly the picture of a volume with a page bearing the inscription "Dedicated to Halina's memory" appeared before my eyes. It was an attractive picture and I clung to it eagerly. But what could I write about? The only thing I knew was physics. If it was to be inscribed to Halina it must not only be a good book, but it must be a book of wide appeal; it must be, therefore, a popular book. A good popular book about modern physics could have this appeal. By this brief chain of argument, starting with a desire to do something, I arrived at this conclusion. I had never tried this sort of writing before, but I thought that I might succeed.

I began to work. I occupied myself with the outline, then with the form of presentation; I wrote three pages systematically every morning, and in the afternoon I planned the pages for the next day. This work brought relief; it strengthened my feeling of life and reality.

I finished the manuscript during my stay in Warsaw. Each evening Halina's father read the result of my daily work and made helpful remarks. In him I had exactly the type of intelligent layman for whom the book was designed. Even Halina's mother—at least at the beginning—approved of my idea, and I succeeded, for brief intervals, in rousing her from her apathy.

The surge of the outside world was brought to me by my work, by the forced contact with publishers, the library, with friends who read the manuscript and made suggestions. Another essential factor was my growing sexual excitement. It was a force in which I felt returning life, only to find out that it pushed me into deeper desperation. By experience I came to understand an ancient, simple truth: physical passion ungoverned by love and merely treated as a necessity is, in the long run, destructive. It leads, through mirages, to greater apathy and to the way down. The sexual excitement strengthened my contact with the outside world but at the same time stained and cheapened my life. It is a period in my life of which I am still unable to write freely, although it formed an essential part of the background of the years following Halina's death. I felt that I should be allowed to do anything which would make me feel more alive and would form new ties binding me to the outside world. Trying to emerge from the circle of despair, I drew others inside the circle. I discovered for the first time the power of sex and took it recklessly, thinking only of myself and damaging the lives of others.

It is destructive to contemplate the past and to nurse a troubled conscience. It is a sign of immaturity. The most sensible attitude is to accept emotionally the inevitability of the past. The attitude of guilt, the attitude of "If only I had not done it," is senseless and undermines our strength of character. It is different with the future. Making our decisions, trying to live and act, we must behave and we do behave as though our wills were free, as though we could decide between good and bad, even when torn by emotions and pricked by desires. It is wise to hold different attitudes toward the past and the future. Looking back on our behavior, we need to believe that we could not have avoided the chain of events. But looking to the future, we must weigh our intentions and plan our actions.

Trying to return to life, I cultivated desires and emotions, nursing them by imagination and self-deceit, not knowing that their source was despair. But the self-deceit was revealed through bitter disappointments, through my own misery and the misery I spread around me by dragging others into my life. Now, after seven years, I still find it hard to understand how I could have deceived myself when the external symptoms of my condition were so unmistakably clear.

XI

Six MONTHS after Halina's death I returned to Lwow. It might have been a different town from that of a year before. It appeared lonely and depressed. I projected my own sadness into the stones, the bricks, the human beings. Loria was touched when he greeted me; he kissed me, as is the Polish custom, and I was deeply moved to see my old friend again.

In previous years Halina had often attended my university lectures. Her place had always been in the fourth row on the right. Even when I was looking elsewhere in the room I could still feel her presence. It was pleasant to know that I could ask her later: "How did you like my lecture? Did you learn something new?"

Now during my first lecture of the term the thought of the

empty place on the right, in the fourth row, darkened my mind; it appeared when I wrote formulae on the blackboard and, pushed brutally away, beat still strongly to the surface of my consciousness. I felt that I was talking gibberish in a dark room, into an empty space, during a never-ending hour.

I tried to avoid our old friends. But I was eager to meet new people whom I had not known before, and especially people outside the periphery of the university. I met some journalists, writers, painters, actors, and found it a relief to see new faces, faces which Halina had not seen. It was a relief to know that they did not associate Halina's name with mine, that for them my tragedy was one of the many stories which they had heard secondhand.

The days were long and burdensome. I had my few hours of lecturing and hours and hours of freedom, which I spent mostly in cafés, reading newspapers over and over, waiting for someone to join me, looking for new acquaintances and avoiding old ones, living an empty, superficial life.

My old friends were disgusted with my behavior. They expected a broken man who would lean on them and to whom they could show their noblest understanding. Instead I did my best to avoid them. Some of my friends understood it, but some of them have never forgiven me for depriving them of a chance to look, free of charge, inside a wounded soul.

I tried to deafen my thoughts with noise and gaiety. But in the back of my mind, while chattering excitedly of things which did not matter, I thought constantly of Halina. There was always the fear that a subject related to my tragedy might enter the conversation. I watched and tried to divert the conversation far from the danger spots. I responded very sensitively, by hysterical gaiety or sudden depression, depending on whether the conversation moved away from or approached these dangerous spots. I tried to weary myself by long talks in stuffy cafés. I never went home before one in the morning, to postpone facing my lonely room and the sleepless night. For years I did not dare to speak the name of Halina aloud.

There was one clear indication of my descent. I did not work at all. I scarcely read scientific papers. For the first time since the opportunity of research was given to me I spent six months without doing any. My paper with Van der Waerden appeared and galley proofs of my book arrived from the Warsaw publisher. Proofreading was the only work I did, but even in this job I was assisted by my chief, Professor C., and by other physicists. I made the final changes in the book and began to grow nervous about its reception. The title was *New Pathways of Science* and it was scheduled for June, for the end of the academic year.

I could not imagine going back to normal life and work. Yet I knew that it was not life, but a spiritless existence that I had chosen; despair imperfectly concealed beneath this empty and superficial existence. I began to weary of the cafés, of the same talks about books, about lectures, Hitler, scandals; of my vulgar flirtations and affairs, of the whole unsavory life, the noise by which I tried to deafen myself. And I realized that I should not be able to emerge by my own strength.

I knew that the first step was to get out of Lwow. Walking through the streets of Lwow, associating everything with Halina, avoiding the house where we had lived, choosing new cafés and acquaintances—all this indicated an inability to face life and weakened my will power until I ceased to believe that I might ever succeed in emerging from the mess. I thought of myself with mingled self-pity and cynicism. The bitterness increased until I decided to talk things over with Professor Loria, who had showed such understanding and sympathy in this difficult time. I said to him:

"I am utterly miserable and I need your help. You brought me here to work and I do nothing. I must get out of Lwow. I hope it is my last request of you. Could you help me to obtain a fellowship? I am sure that leaving Lwow is what I need."

Loria was sympathetic.

"I have thought about it myself. We could try a Rockefeller fellowship. You are still a few months under thirty-five, which is the age limit. Either you get it this year or never."

During our next meetings we discussed the problem in detail.

Loria told me of an essential condition for becoming a fellow of the Rockefeller Foundation:

"What the Rockefeller Foundation wants to know about you before they give you their holy money is whether you can go further. They have the attitude of good American businessmen. They want to invest their money well. I must be able to assure them that sooner or later you will get a professorship in Poland. I believe I can do it. Your noble Master N. in Cracow will have to retire soon. Either you will get a professorship in Cracow or, after some shifting, they will have to fill the remaining hole with you. You are practically the only docent in theoretical physics. I hope that the anti-Semitism here will decline. They must see what it brought to Germany. You stand a good chance, I am sure, and the Foundation may take my word for it."

The application for a Rockefeller fellowship must be signed by two professors. The more important the names, the better. It was obvious that Loria would be my sponsor. We considered whom to ask to support the application.

There was one danger. Usually one physicist or mathematician left Poland each year for a Rockefeller fellowship. Professor P., head of the department in experimental physics in Warsaw, was at that time most influential at the Foundation and his opinion carried great weight. If he decided to sponsor someone else, I should not stand a chance. Therefore the plan which I suggested was obvious:

"It would be a good thing if Professor P. from Warsaw would support my application. It would increase my chances immensely."

Loria liked the idea and wrote to Professor P. He agreed to sign the application. Aside from the fact that he was one of the more liberal professors at the university, I am sure he was moved in part by my tragedy, which was well known in the world of Polish physics and had awakened universal sympathy. Even in death Halina helped me in my struggle.

There was one question which had yet to be settled before the application was finally made out: where to go. I discussed the problem with Loria. I reasoned:

"I must show knowledge of the language of the country to which I intend to go. This restricts the possibilities. England is out because I speak not a word of English. Adolf's Germany is out. There is Zurich, with excellent theoretical physics; perhaps Paris, as I talk a little French and could easily gain some fluency; then Leyden in Holland or Copenhagen, where everybody at the university speaks German too. It is difficult to choose. What would you advise?"

"I should advise you to go somewhere quite different. You say you cannot go to England because you do not know English. I should say quite the opposite. You should go to England for that very reason. This is exactly what I did years ago, and it opened new vistas in my life. You will learn any language if necessary. And, what is more important, you will come in contact with teaching, with ways of thinking and people entirely different from everything you know. If I were you I should go to England and start learning English tomorrow or, still better, today. And when I say England I mean Cambridge, where there is perhaps the best physics in the world."

Loria convinced me. I had a new occupation and an excuse for not doing scientific work: I must learn English.

The preliminary formalities connected with the Rockefeller fellowship were arranged without difficulty. According to the rules, each candidate must pass the scrutiny of an officer of the Foundation. Dr M. from the Paris office came to Lwow and asked me about my plans. I explained in German that it would be my first long trip to a place with a scientific atmosphere. He seemed impressed that even the short trip to Leipzig the year before had been enough to produce some results. With Loria he talked mostly about my prospects. Concerning the language difficulties, he was satisfied by my assurance that I planned to go to England two months in advance, at my own expense, to practice the language and that I was already studying it. Then a doctor examined me and inoculated me against typhoid. In May I got the fellowship to go to England for a year. The payment was thirty pounds a month and traveling expenses in both directions.

In June *New Pathways of Science* appeared with a page on

which was printed "Dedicated to Halina's memory." The book was an immediate success, was displayed in every bookshop window, was widely read and had very good reviews. I got the first taste of fame, but before I could enjoy its full flavor I had to go to England, leaving my local prestige behind me.

Before I left Poland I visited Halina's parents. Her mother was much improved; her face was much older but quieter, with new lines sculptured by tragedy. She tried, at least on the surface, to lead a normal life. My parents-in-law bought a house outside Warsaw which they furnished with our Lwow furniture. They wanted to breathe the atmosphere of the past but did not have the courage to hang even one picture of Halina.

I came to them in my black tie, with a black band around my arm, as the custom is in Poland. I wore it as a safeguard against personal questions; it made people more cautious in their conversation.

While staying with my parents-in-law I had a feeling of guilt. I thought:

"Time has loosened the bonds between us and will make them still looser. I am now plunging for the last time into the memories of my past before leaving the Continent. In my suitcase I have a new light gray suit, a new blue tie which I shall wear on the first day of my journey. They don't know about this. In England no one has heard about my tragedy, and I shall do my best to hide and to forget it. The sorrow which I am showing now is perhaps real, but in its external display it is a deliberate performance."

England is far from Poland. It is not only a thousand miles away; it is centuries distant. For me, in my childhood surroundings, England was as far as the moon. This attitude remained. The journey was a great adventure, nearly a miracle, an entry into a new world. My life was not finished. I felt the tense expectation of tomorrow awakening in me once more.

XII

Cambridge is an isolated island. Erasmus of Rotterdam, coming out of his grave and walking through the streets, would be mildly surprised by a few changes, but he would still recognize the place where he taught and he would still sniff the scent of learning saturated with centuries of tradition. It was for me a new, strange world, entirely different from the Continent.

Students in Poland and Germany had complete freedom. They could come to lectures or not, as they pleased. Many of them lived in great poverty, always trying to get small jobs and earn enough money to pursue their studies.

In Cambridge students of the same age are treated like privileged adolescents. They live luxuriously in the old colleges, each student having two comfortable rooms and a bathroom. Each evening, in his student's gown, he attends an impressive dinner in his college hall. On the street his behavior is watched by a "proctor," followed at a discreet distance by two "bulldogs," in top hats, who enforce the laws of the colleges. The college authorities even watch the sex lives of the students and have the right to expel prostitutes from the town. Each student has a supervisor and a tutor. To his supervisor he reports his progress in studies and receives from him instructions for further work. His tutor fills a more personal role: guards his morals; functions, as it were, *in loco parentis*, giving permission to his chaperoned girl friend to visit the student on Sunday.

In Poland the majority of students, coming mostly from the anti-Semitic middle class, were the stronghold of reaction. It was at the Polish universities that anti-Semitic and reactionary slogans sank in most deeply and from there they spread over the country, cooked in the sauce of patriotic phraseology. The majority of students fought liberalism with cheap patriotic phrases behind which they concealed their hatred of everything which might lead to social progress.

It is different in Cambridge. Here youth is more progressive than its parents. Here I witnessed a student pacifist demonstration, contrasting with the memory of the noisy demonstrations of students in my country who shouted slogans, urging hatred, war and the extermination of Jews. Here in their debates, modeled on the procedure of the English Parliament, progressive resolutions usually won. I remember the students' discussion on a motion: *This house has more faith in Roosevelt's than in Baldwin's policy,* which passed by a majority. The sons of the English Tories relax in Cambridge and furbish their consciences for their future. Progressive and even radical in college, they prepare to serve the British Empire later with the wisdom gained from this radical past.

In Cambridge a college is a building in which the students live, eat and are supervised. These colleges are built in the austere English Gothic style and have quadrangles covered with thick, carefully cut grass. Cambridge is an isolated system like a thermos bottle or a ghetto. The isolating factor is the tradition of centuries which has darkened the red bricks of Trinity Gate and the white stones of King's Chapel.

The university is something different from the colleges. It gives the official lectures and degrees. But to be admitted to the university one must be a member of a college. Whenever I was asked what I was doing in England and I answered: "I am in Cambridge on a Rockefeller fellowship," I always heard the same question:

"To what college are you attached?"

It was difficult to make clear to an Englishman my peculiar position. I explained:

"I am not attached to any college. I am privately subsidized by the Rockefeller Foundation to work on some problems with certain professors. I am completely free and not subject to any college discipline. I may go out at two o'clock in the morning and I may invite any girl friends I like to call on me. I don't wear a gown, but I am allowed to visit any lectures I please. My only formal obligation is to stay in Cambridge, my only moral obligation is to work."

I had the best intentions of fulfilling my moral obligation.

The greatest theoretical physicist in Cambridge was P. A. M. Dirac, one of the outstanding scientists of our generation, then a young man about thirty. He still occupies the chair of mathematics, the genealogy of which can be traced directly to Newton.

I knew nothing of Dirac, except that he was a great mathematical physicist. His papers, appearing chiefly in the *Proceedings of the Royal Society*, were written with wonderful clarity and great imagination. His name is usually linked with those of Heisenberg and Schroedinger as the creators of quantum mechanics. Dirac's book *The Principles of Quantum Mechanics* is regarded as the bible of modern physics. It is deep, simple, lucid and original. It can only be compared in its importance and maturity to Newton's *Principia*. Admired by everyone as a genius, as a great star in the firmament of English physics, he created a legend around him. His thin figure with its long hands, walking in heat and cold without overcoat or hat, was a familiar one to Cambridge students. His loneliness and shyness were famous among physicists. Only a few men could penetrate his solitude. One of the fellows, a well-known physicist, told me:

"I still find it very difficult to talk with Dirac. If I need his advice I try to formulate my question as briefly as possible. He looks for five minutes at the ceiling, five minutes at the windows, and then says 'Yes' or 'No.' And he is always right."

Once—according to a story which I heard—Dirac was lecturing in the United States and the chairman called for questions after the lecture. One of the audience said:

"I did not understand this and this in your arguments."

Dirac sat quietly, as though the man had not spoken. A disagreeable silence ensued, and the chairman turned to Dirac uncertainly:

"Would you not be kind enough, Professor Dirac, to answer this question?"

To which Dirac replied: "It was not a question; it was a statement."

Another story also refers to his stay in the United States. He lived in an apartment with a famous French physicist and they invariably talked English to each other. Once the French physicist, finding it difficult to explain something in English, asked Dirac, who is half English and half French:

"Do you speak French?"

"Yes. My mother was French," answered Dirac in an unusually long sentence. The French professor burst out:

"And you say this to me now, having allowed me to speak my bad, painful English for weeks! Why did you not tell me this before?"

"You did not ask me before," was Dirac's answer.

But a few scientists who knew Dirac better, who managed after years of acquaintance to talk to him, were full of praise of his gentle attitude toward everyone. They believed that his solitude was a result of shyness and could be broken in time by careful aggressiveness and persistence.

These idiosyncrasies made it difficult to work with Dirac. The result has been that Dirac has not created a school by personal contact. He has created a school by his papers, by his book, but not by collaboration. He is one of the very few scientists who could work even on a lonely island if he had a library and could perhaps even do without books and journals.

When I visited Dirac for the first time I did not know how difficult it was to talk to him as I did not then know anyone who could have warned me.

I went along the narrow wooden stairs in St John's College and knocked at the door of Dirac's room. He opened it silently and with a friendly gesture indicated an armchair. I sat down and waited for Dirac to start the conversation. Complete silence. I began by warning my host that I spoke very little English. A friendly smile but again no answer. I had to go further:

"I talked with Professor Fowler. He told me that I am supposed to work with you. He suggested that I work on the internal conversion effect of positrons."

No answer. I waited for some time and tried a direct question: "Do you have any objection to my working on this subject?" "No."

At least I had got a word out of Dirac.

Then I spoke of the problem, took out my pen in order to write a formula. Without saying a word Dirac got up and brought paper. But my pen refused to write. Silently Dirac took out his pencil and handed it to me. Again I asked him a direct question to which I received an answer in five words which took me two days to digest. The conversation was finished. I made an attempt to prolong it.

"Do you mind if I bother you sometimes when I come across difficulties?"

"No."

I left Dirac's room, surprised and depressed. He was not forbidding, and I should have had no disagreeable feeling had I known what everyone in Cambridge knew. If he seemed peculiar to Englishmen, how much more so he seemed to a Pole who had polished his smooth tongue in Lwow cafés! One of Dirac's principles is:

"One must not start a sentence before one knows how to finish it."

Someone in Cambridge generalized this ironically:

"One must not start a life before one knows how to finish it."

It is difficult to make friends in England. The process is slow and it takes time for one to graduate from pleasantries about the weather to personal themes. But for me it was exactly right. I was safe because nobody on the island would suddenly ask me: "Have you been married?" No conversation would even approach my personal problems. The gossipy atmosphere of Lwow's cafés belonged to the past. How we worked for hours, analyzing the actions and reactions of others, inventing talks and situations, imitating their voices, mocking their weaknesses, lifting gossip to an art and cultivating it for its own sake! I was glad of an end to these pleasures. The only remarks which one is likely to hear from an Englishman, on the subject of another's personality, are:

"He is very nice."
"He is quite nice."
Or, in the worst case:
"I believe that he is all right."

From these few variations, but much more from the subtle way in which they are spoken, one can gain a very fair picture after some practice. But the poverty of words kills the conversation after two minutes.

The first month I met scarcely anyone. The problem on which I worked required tedious calculations rather than a search for new ideas. I had never enjoyed this kind of work, but I determined to learn its technique. I worked hard. In the morning I went to a small dusty library in the Cavendish Laboratory. Every time I entered this building I became sentimental. If someone had asked me, "What is the most important place in the world?" I would have answered: "The Cavendish Laboratory." Here Maxwell and J. J. Thomson worked. From here, in the last years under Rutherford's leadership, ideas and experiments emerged which changed our picture of the external world. Nearly all the great physicists of the world have lectured in this shabby old auditorium which is, by the way, the worst I have ever seen.

I studied hard all day until late at night, interrupted only by a movie which took the place of the missing English conversation. I knew that I must bring results back to Poland. I knew what happened to anyone who returned empty-handed after a year on a fellowship. I had heard conversations on the subject and I needed only to change the names about to have a complete picture:

A: I saw Infeld today; he is back already. What did he do in England?
B: We have just searched carefully through the science abstracts. He didn't publish anything during the whole year.
A: What? He couldn't squeeze out even one brief paper in twelve months, when he had nothing else to do and had the best help in the world?
B: I'm sure he didn't. He is finished now. I am really very sorry for him. Loria ought to have known better than to make a fool of himself by recommending Infeld for a Rockefeller fellowship.

A: We can have fun when Loria comes here. We'll ask him what his protégé did in England. Loria is very talkative. Let's give him a good opportunity.

B: Yes. It will be quite amusing. What about innocently asking Infeld to give a lecture about Cambridge and his work there? It will be fun to see him dodging the subject of his own work.

This is the way academic failure was discussed in Poland. I should have little right to object. Bitter competition and lack of opportunity create this atmosphere.

When I came to Cambridge, before the academic year began, I learned that Professor Born would lecture there for a year. His name, too, is well known to every physicist. He was as famous for the distinguished work which he did in theoretical physics as for the school which he created. Born was a professor in Goettingen, the strongest mathematical center of the world before it was destroyed by Hitler. Many mathematicians and physicists from all over the world went to Goettingen to do research in the place associated with the shining names of Gauss in the past and Hilbert in the present. Dirac had had a fellowship in Goettingen and Heisenberg obtained his docentship there. Some of the most important papers in quantum mechanics were written in collaboration by Born and Heisenberg. Born was the first to present the probability interpretation of quantum mechanics, introducing ideas which penetrated deeply into philosophy and are linked with the much-discussed problem of determinism and indeterminism.

I also knew that Born had recently published an interesting note in *Nature,* concerning the generalization of Maxwell's theory of electricity, and had announced a paper, dealing at length with this problem which would appear shortly in the *Proceedings of the Royal Society.*

Being of Jewish blood, Professor Born had to leave Germany and immediately received five offers, from which he chose the invitation to Cambridge. For the first term he announced a course on the theory on which he was working.

I attended his lectures. The audience consisted of graduate students and fellows from other colleges, chiefly research workers. Born spoke English with a heavy German accent. He was about fifty, with gray hair and a tense, intelligent face with eyes in which the suffering expression was intensified by fatigue. In the beginning I did not understand his lectures fully. The whole general theory seemed to be sketchy, a program rather than a finished piece of work.

His lectures and papers revealed the difference between the German and English style in scientific work, as far as general comparisons of this kind make any sense at all. It was in the tradition of the German school to publish results quickly. Papers appeared in German journals six weeks after they were sent to the editor. Characteristic of this spirit of competition and priority quarrels was a story which Loria told me of a professor of his in Germany, a most distinguished man. This professor had attacked someone's work, and it turned out that he had read the paper too quickly; his attack was unjustified, and he simply had not taken the trouble to understand what the author said. When this was pointed out to him he was genuinely sorry that he had published a paper containing a severe and unjust criticism. But he consoled himself with the remark: "Better a wrong paper than no paper at all."

The English style of work is quieter and more dignified. No one is interested in quick publishing, and it matters much less to an Englishman when someone else achieves the same results and publishes them a few days earlier. It takes six months to print a paper in the *Proceedings of the Royal Society*. Priority quarrels and stealing of ideas are practically unknown in England. The attitude is: "Better no paper at all than a wrong paper."

In the beginning, as I have said, I was not greatly impressed with Born's results. But later, when he came to the concrete problem of generalizing Maxwell's equations, I found the subject exciting, closely related to the problems on which I had worked before. In general terms the idea was:

Maxwell's theory is the theory of the electromagnetic field, and it forms one of the most important chapters in theoretical physics. Its great achievement lies in the introduction of the concept of the *field*. It explains a wide region of experimental facts but, like every theory, it has its limitations. Maxwell's theory does not explain why elementary particles like electrons exist, and it does not bind the properties of the field to those of matter.

After the discovery of elementary particles it was clear that Maxwell's theory, like all our theories, captures only part of the truth. And again, as always in physics, attempts were made to cover, through modifications and generalizations, a wider range of facts. Born succeeded in generalizing Maxwell's equations and replacing them by new ones. As their first approximation these new equations gave the old laws confirmed by experiments. But in addition they gave a new solution representing an elementary particle, the electron. Its physical properties were determined to some extent by the new laws governing the field. The aim of this new theory was to form a bridge between two hitherto isolated and unreconciled concepts: field and matter. Born called it the Unitary Field Theory, the name indicating the union of these two fundamental concepts.

After one of his lectures I asked Born whether he would lend me a copy of his manuscript. He gave it to me with the assurance that he would be very happy if I would help him. I wanted to understand a point which had not been clear to me during the lecture and which seemed to me to be an essential step. Born's new theory allowed the construction of an elementary particle, the electron, with a finite mass. Here lay the essential difference between Born's new and Maxwell's old theories. A whole chain of argument led to this theoretical determination of the mass of the electron. I suspected that something was wrong in this derivation. On the evening of the day I received the paper the point suddenly became clear to me. I knew that the mass of the electron was wrongly evaluated in Born's paper and I knew how to find the right value. My whole argument seemed simple and convincing to me. I could hardly wait to tell it to Born, sure that he would see my point immediately. The next day I went to him after his lecture and said:

"I read your paper; the mass of the electron is wrong."

Born's face looked even more tense than usual. He said:

"This is very interesting. Show me why."

Two of his audience were still present in the lecture room. I took a piece of chalk and wrote a relativistic formula for the mass density. Born interrupted me angrily:

"This problem has nothing to do with relativity theory. I don't like such a formal approach. I find nothing wrong with the way I introduced the mass." Then he turned toward the two students who were listening to our stormy discussion.

"What do you think of my derivation?"

They nodded their heads in full approval. I put down the piece of chalk and did not even try to defend my point.

Born felt a little uneasy. Leaving the lecture room, he said:

"I shall think it over."

I was annoyed at Born's behavior as well as at my own and was, for one afternoon, disgusted with Cambridge. I thought: "Here I met two great physicists. One of them does not talk. I could as easily read his papers in Poland as here. The other talks, but he is rude." I scrutinized my argument carefully but could find nothing wrong with it. I made some further progress and found that new and interesting consequences could be drawn if the "free densities" were introduced relativistically. A different interpretation of the unitary theory could be achieved which would deepen its physical meaning.

The next day I went again to Born's lecture. He stood at the door before the lecture room. When I passed him he said to me:

"I am waiting for you. You were quite right. We will talk it over after the lecture. You must not mind my being rude. Everyone who has worked with me knows it. I have a resistance against accepting something from outside. I get angry and swear but always accept after a time if it is right."

Our collaboration had begun with a quarrel, but a day later complete peace and understanding had been restored. I told Born about my new interpretation connecting more closely and clearly, through the "free densities," the field and particle aspects. He immediately accepted these ideas with enthusiasm. Our collaboration grew closer. We discussed, worked together after lectures, in Born's home or mine. Soon our relationship became informal and friendly.

I ceased to work on my old problem. After three months of my stay in Cambridge we published together two notes in *Nature*, and a long paper, in which the foundations of the New Unitary Field Theory were laid down more deeply and carefully than before, was ready for publication in the *Proceedings of the Royal Society*.

For the first time in my life I had close contact with a famous, distinguished physicist, and I learned much through our relationship. Born came to my home on his bicycle whenever he wished to communicate with me, and I visited him, unannounced, whenever I felt like it. The atmosphere of his home was a combination of high intellectual level with heavy Germany pedantry. In the hall there was a wooden gadget announcing which of the members of the family were out and which were in.

I marveled at the way in which he managed his heavy correspondence, answering letters with incredible dispatch, at the same time looking through scientific papers. His tremendous collection of reprints was well ordered; even the reprints from cranks and lunatics were kept, under the heading "Idiots." Born functioned like an entire institution, combining vivid imagination with splendid organization. He worked quickly and in a restless mood. As in the case of nearly all scientists, not only the result was important but the fact that he had achieved it. This is human, and scientists are human. The only scientist I have ever met for whom this personal aspect of work is of no concern at all is Einstein. Perhaps to find complete freedom from human weakness we must look up to the highest level achieved by the human race. There was something childish and attractive in Born's eagerness to go ahead quickly, in his restlessness and his moods, which changed suddenly from high enthusiasm to deep depression. Sometimes when I would come with a new idea he would say rudely, "I think it is rubbish," but he never minded if I applied the same phrase to some of his ideas. But the great, the celebrated Born was as happy and as pleased as a young student at words of praise and encouragement. In his enthusiastic attitude, in the vividness of his mind, the impulsiveness with which he grasped and rejected ideas, lay his great charm. Near his bed he had always a pencil and a piece of paper on which to scribble his inspirations, to avoid turning them over and over in his mind during sleepless nights.

Once I asked Born how he came to study theoretical physics. I was interested to know at what age the first impulse to choose a definite path in life crystalizes. Born told me his story. His father was a medical man, a university professor, famous and rich. When he died he left his son plenty of money and good advice. The money was sufficient, in normal times, to assure his son's independence. The advice was simply to listen during his first student year to many lectures on many subjects and to make a choice only at the end of the first year. So young Born went to the university at Breslau, listened to lectures on law, literature, biology, music, economics, astronomy. He liked the astronomy lectures the most. Perhaps not so much for the lectures themselves as for the old Gothic building in which they were held. But he soon discovered that to understand astronomy one must know mathematics. He asked where the best mathematicians in the world were to be found and was told "Goettingen." So he went to Goettingen, where he finished his studies as a theoretical physicist, habilitated and finally became a professor.

"At that time, before the war," he added, "I could have done whatever I wanted with my life since I did not even know what the struggle for existence meant. I believe I could have become a successful writer or a pianist. But I found the work in theoretical physics more pleasant and more exciting than anything else."

Through our work I gained confidence in myself, a confidence that was strengthened by Born's assurance that ours was one of the pleasantest collaborations he had ever known. Loyally he stressed my contributions in his lectures and pointed out my share in our collaboration. I was happy in the excitement of obtaining new results and in the conviction that I was working on essential problems, the importance of which I certainly exaggerated. Having new ideas, turning blankness into understanding, suddenly finding the right solution after weeks or months of painful doubt, creates perhaps the highest emotion man can experience. Every scientist knows this feeling of ecstasy even if his achievements are small. But this pure feeling of *Eureka* is mixed with overtones of very human, selfish emotions: "*I* found it; *I* will have an important paper; it will help me in my career." I was fully aware of the presence of these overtones in my own consciousness.

Our papers were quoted, discussed and read. After three months in Cambridge I had achieved results. I thought, "Even if I do nothing more, I have made a success of my fellowship."

I should like to stress one point clearly. The essential work in theoretical physics lies in formulating new and basic ideas. In the work we were doing this had been largely done before by Born. What we did together was to deepen and broaden, to generalize and develop these ideas. Although the whole theory has usually been associated with both our names, the first paper referring to this subject was written by Born before we even met.

The theory on which I was working with Born was classical in its aspect, and this classical aspect of our theory was presented fairly consistently and completely in our paper. The next problem was to generalize these ideas, to go one step higher, to leave behind the classical theories which break down for the interior of the atom, where the quantum theories reign. We followed only the first part of the pattern by formulating the classical unitary field theory. The next step, which we knew would be much more difficult, was to find the quantum mechanical generalization. Born began immediately to burn with plans. He reminded me of an author who had got in the habit of writing ten pages of his novel daily and, having written the last six pages of one novel, started to write down four pages of a new one. I did not like Born's new ideas, and at that time I was little interested in the quantum mechanical aspect of our problem. The restlessness and speed with which Born worked forced him to lose perspective on his own work, which can be gained only by the proper time distance. I did not believe that anything important would come out of Born's new attempts.

Knowing that I had saved myself from possible disgrace on my return to Poland, I decided to take a rest. Everything in England was new and different. God only knew when I should be as free again. I wanted to learn something of the country, to savor it slowly and to find its taste. I listened politely as Born explained his new theories but I did nothing about them. Born worked for nearly two months on the quantum mechanics but nothing came out. He was probably too tired but did not want to admit it. I tried to live in leisurely fashion and looked for emotions and pleasures outside physics, even outside Cambridge.

The fact that I was able to work again after my tragedy, and on a level not achieved before, made me drunk with self-confidence. For the first time since Halina's death I had a feeling of relaxation. England became for me the most wonderful place in the world. Just to exist, to walk through the streets looking at the bricks and faces, repeat to myself, "I am in England," was enough to create a warm feeling of satisfaction.

I was brought up in a country where a Jew going through the streets outside his ghetto had the feeling, consciously or subconsciously, "These are my enemies." In this hostile atmosphere the slightest provocation might cause a fight. The policeman was harsh and the government official, underpaid and browbeaten by his superiors, was drunk with the power of his office. There were taboos which must not be violated. The honor of the soldier's uniform was one of them. A smiling, friendly face was seldom seen. Everyone was regarded as an enemy until proved the contrary.

I remembered my journeys in Polish railways. In the small compartments accidental cross sections of the Polish population were created and people talked freely to each other. The conversation might start on any subject—the weather, vacations, politics—but sooner or later it settled on the inevitable topic and there it would remain: the Jewish problem. I remembered my journey to England and the last sounds of my native language. I sat silently in the corner looking at the newspaper and dreaming of England. A fresh-faced young girl and her mother, a priest, a rich farmer and a smart officer were talking. My cheeks burning with hate and shame, I tried to ignore them, to shut them out of my consciousness.

"You are fortunate in Poznan. You have very few Jews there."

"Still too many for my taste."

"Poland is for Poles."

"Jews ought to go to Palestine."

"Hitler is clever. He taught them a lesson."

"We need a Hitler. A Polish Hitler is what we need."

"Our government is too patient with Jews."

"They bring the infection of Bolshevism."

"Jews are Bolsheviks."

"Jews are ruining the country."

"I knew one decent Jew."

"I never buy at a Jewish shop."

"We ought to learn from Germany."

"Jews and Communists—these are our dangers."

Once I asked a direction of a policeman in Cambridge. He took my arm, and my timid foreigner's soul was frightened. He helped me across the street, accompanied me half way and expressed regret that he could not go further. He hoped that I would have a nice walk and analyzed the weather. I felt friendliness and security everywhere around me. Somebody asked me, "What do you like most in England?" and I answered, seriously believing my words:

"The conviction that if I go through the streets of an English town and step on someone's tender corn I will hear an apology: 'I am so sorry.' "

I well knew that the friendliness, the subtlety, the curious sense of humor and the courage of the Englishman came from complete detachment and I also knew how annoying this could be. In the English, emotions and passions never rise to the surface. A strange people, living on an island, eating the worst food in the world and trying to keep the Empire together, they do everything without fire. Handsome, well-dressed men reading *The London Times* in the train retain their bland indifference when a beautiful woman enters the compartment. What does a nation do which does not care about food and sex? The answer is provided by English history: it builds an empire.

My state, my past tragedy, all that England did for me awakened my admiration for that country. In my condition of oversensitiveness after Halina's death I considered myself placed in a great sanatorium especially designed for me and subsidized by the Rockefeller Foundation. To be in England, to live, to work, to die in Cambridge was for me a pattern of the highest happiness man can achieve.

At that time my popular book was translated into English and published under the title *The World in Modern Science*. At the publisher's suggestion I asked Einstein for an introduction. I received a letter praising my book highly and a brief but charming introduction. Einstein was even kind enough to remark in his letter that he stood ready to change his preface if it were not suitable. The book was a moderate success and some of the reviews, especially that in *Nature*, were extremely good.

In the spring I felt a desire to return to scientific work. I attended Dirac's lectures on quantum electrodynamics. It was the problem of forming a bridge from Maxwell's theory to the quantum theory of the electromagnetic field; one of the problems concerned the transition from classical physics to quantum physics, to physics of the atom. Dirac did it, as he did everything, in an original, elegant manner. Thinking over the lecture, it occurred to me that a similar procedure could be applied to the theory suggested by Born and developed by us, i.e., the unitary field theory.

The unitary field theory is a generalization of Maxwell's classical electrodynamics. We were then searching for a transition into the field of quantum physics. I found that the steps leading from Maxwell's theory to quantum electrodynamics, as presented by Dirac, suggested the way which we were seeking. I worked out the initial stages of the mathematical scheme and showed the results to Born. He liked the idea mildly but pointed out its provisional character. For the next part of my fellowship we worked together on this problem, producing two long papers and two short notes. The result of our collaboration was formulated in three long papers and four brief notes. In the scientific world in Poland, where the number of papers is carefully counted and weighed, I felt sure that seven publications in one year would be considered an impressive accomplishment. Although I do not now rate the theory as highly as I did at that time, and although in the end the theory proved less successful than we had thought, I still believe that these papers are valuable since they cleared up the problem of generalizing Maxwell's equations and presented a form of transition from classical electrodynamics to quantum electrodynamics which was later used even in problems of different character.

I was happy in England and wanted to stay longer. Born was also anxious to prolong our collaboration, and the Rockefeller Foundation granted me an additional three months. It seemed from the external signs that I had left the worst struggle behind me and that at thirty-six my future might soon be crystallized by a permanent professorship.

At the end of the year which I spent in England Loria visited me in Cambridge. He had been in England for two years just before the World War, and he wanted to refresh the friendships and acquaintances made at that time. He took a room in the house where I lived, and although I was very happy to see my old friend, our talks brought up my return to Poland and made me realize that my stay in Cambridge was ending.

We sat on a bench on the lawn behind Trinity College with its bright flowers and thick grass and its trees reflected in the small picturesque river flowing behind the old colleges. In these peaceful surroundings Loria began to talk seriously of my future.

"Anti-Semitism is slowly but steadily growing in Poland. Events in Germany have had their influence upon Poland, and I only hope that this influence will not increase."

As always, Loria spoke in rounded, polished sentences, strongly emphasizing the important words. "Anti-Semitism" was especially emphasized. I had forgotten, or rather had tried to forget, its existence. Was not this one of the essential reasons why I had felt so happy in Cambridge? Never in my life had I breathed air so devoid of the germs of race hatred as in Cambridge. But my life in England was ending. I had to think of tomorrow, and for tomorrow my friend brought me the message:

"Anti-Semitism is slowly but steadily growing in Poland."

All my plans for the future depended on whether or not this growth would continue. Loria proceeded:

"Now, after your work and the results achieved, it will be difficult not to give you a chair. Professor N., your great master, finally retired, ironically enough, with you as his only pupil. Probably they will do some shifting; the professor from Wilno may go to Cracow. That may take some time. But I don't see how they can offer the chair in Wilno to anyone but you. There is still not one 'Aryan' docent in theoretical physics in Poland, and this is your chance."

I expressed my doubts:

"But you know, Professor Loria, that the universities are most ingenious in this respect and they can play tricks which neither of us can foresee at this moment. As a matter of fact I don't believe very seriously myself that they will succeed, but they may try."

Loria did not remove my doubts.

"I am sure of this. I have heard rumors already. The idea is not to fill the empty chair but to wait for some time until a new 'pure' theoretical physicist pops out. But even then the situation is not hopeless, because there is a possibility that Lwow will get a new chair in theoretical mechanics. Although, on the other hand, even if the faculty is granted the chair, it can say that you are not a specialist in mechanics and select someone else."

I began to lose track of all the possible combinations affecting my future. I was sick at the thought of having to contend once more with all the intrigue and hypocrisy. The year in Cambridge had made me conceited and impatient. How could it make such a difference that I was a Jew? What difference did it really make?

XIII

I WENT BACK TO LWOW. Return is always difficult. The characteristics of a provincial town and of provincial life, the lack of proportion, were annoying instead of amusing. Lwow considered its problems as though Lwow were the center of all the world. And I knew too well by that time that Cambridge, and more specifically the Cavendish Laboratory, was the center of the world. I nursed a grievance against all Poles who didn't behave like English gentlemen, who didn't smile nicely, who didn't talk about the weather and who didn't eat bacon and eggs for breakfast. I had contempt for cafés and decided to propagate the English style of life in Poland, until after a few months I found myself sitting again in cafés. I still worked on the unitary field theory, corresponded with Born and dreamed of the Trinity Gate, planning to spend my next vacation in England.

A new biennial meeting of the Polish Physical Society was announced, this time in my native town of Cracow. The meetings are vanity markets where everyone exposes the stock he has collected during the past two years in a ten-minute lecture to which hardly anyone listens. The only interesting things about the meetings are the discussions, talks and gossip going on outside the lecture room.

Wilno, Poznan, Warsaw, Cracow—four meetings in six years —marking, in two-year intervals, important changes in my life. At the time of the first meeting I was a teacher in a Jewish gymnasium. In Wilno I had met Halina. In Poznan I had felt that I would have to leave Polish academic life. I was not interested in the Warsaw meeting; it had come too soon after Halina's death. Now Cracow, where I had been born and had gone to school.

I was invited to give an hour lecture on the unitary field theory. This was in itself a distinction, since the one-hour lectures differed in character from the ten-minute routine reports. The hour lectures were held not in sections but for all members of the society, and were more general in character. In them skill of exposition was as important as subject matter.

At that time I knew that my only chance of gaining a professorship was at the university of Wilno. Professor W. from Wilno had been invited to Cracow, and his former chair remained unoccupied. Loria said to me:

"Your lecture at the meeting is extremely important. All the physicists from Wilno will listen carefully. I should like them to see what a well-thought-out and interestingly delivered lecture is like. I intend to talk with them later about you and to let them read the letter which I received from the Rockefeller Foundation. Mr Rockefeller is very much impressed by the work you did in Cambridge and asks me what your chances are for a professorship in Wilno. You may rest assured that I shall try my best to play all these cards well."

In the big auditorium, crowded by physicists and visitors, I lectured at Cracow's Physical Society meeting. It was in the same auditorium where, sixteen years before, I had sat as a student. I was invited to speak for another hour to a more select audience. My talk was a great success. One of the professors told me afterward:

"Listening to the way you presented the problem was a rich scientific experience which I shall not forget."

While I awaited developments a new success and new opportunities for earning arose. My popular book became widely read in Poland. The former minister of finance, Mr Matuszewski, read it. He was one of the most cultured and intelligent men in the government and was known to be a friend of Marshall Pilsudski. At that time Matuszewski had left his post as a cabinet minister and assumed the editorship of a new daily paper called *Gazeta Polska*, subsidized by the government and of course supporting the government. The paper had a high intellectual level and printed many articles devoted to literature and science. It was Matuszewski's ambition to collect around the paper the best names and pens in Poland. Unexpectedly I received a charming letter from him, telling me how delightful he had found my book and inviting me to be a free-lance correspondent. He left it to me to decide how much and how often I intended to write and asked me to state my terms.

I had some hesitation about accepting the offer because of the political situation. At that time Pilsudski was still alive. The new government party which had come into power in 1926 through a revolution was, in the beginning, supported by all progressive elements, by workers and peasants and even by the outlawed Communist party. The refusal of railway workers to transport fighting troops for the old administration had been a decisive factor in Pilsudski's victory. It was, in the beginning, a fight between the progressive elements and the reactionary "Polish National Democracy" which had proclaimed slogans similar to those by which Hitler took power in Germany.

Slowly but steadily the new government moved away from its progressive plans until, deserted by peasants and workers, it became a small party, lacking the support of the people and pushed by events toward autocracy. The only merit which still remained was that the government prevented the openly fascist "National Democracy" from taking power, that at least in fighting this party the government had sometimes the courage to fight anti-Semitism and reactionary slogans. The choice at this time was between the mild fascism of the government and the reckless, brutal fascism of the National Democrats.

In these circumstances, I thought, I might well accept the offer to become one of the few correspondents of the government paper which stood for liberal thought. Matuszewski, whom I met and who was a talented writer, represented the more liberal wing of the party. I was impressed by the clarity of his thoughts and his subtle intelligence. (As I learned later from the newspapers, it was he who was responsible for the successful transfer of gold from Poland to Paris, following the German invasion.)

I was very well paid. The four articles which I wrote each month for the paper brought me more than my university salary. I had never thought that money could be earned so easily. Whenever I could, while writing about physics, education or other subjects, I tried to show the destruction of culture which was among fascism's gifts to Germany, and I even directly attacked the rising anti-Semitism in the Polish universities. All my articles were printed and my name became well known in Poland. The experience with the *Gazeta Polska* was exciting and pleasant. I liked the taste of hasty newspaper work, its Bohemian atmosphere and the consideration which I received everywhere by accumulation of two titles: docent, and special correspondent of a government paper.

As I stated before, the government party was mildly anti-Semitic. This statement requires some explanation. The government party was not anti-Semitic in its feeling, if such a general phrase makes any sense at all. The party was, however, anti-Semitic in its official actions, trying, in its policy and economy, to distinguish between the interests of the Jewish and non-Jewish population. The Jewish problem had always existed in Poland. It was the problem of Polish towns overcrowded by Jews and of Polish commerce largely in the hands of Jews. The division of all human beings into two classes—Jews and non-Jews—was deeply rooted in the psychology of every Pole. Only a very thin layer of the radical Polish intelligentsia was free of this Jewish complex, and even they achieved it in a rather perverse way, by acclaiming the Jew as the salt of the earth and highly exaggerating his virtues. The attitude of these radicals showed a prejudice in the opposite direction and indicated that they too accepted the division even though they drew opposite conclusions.

Superficially and schematically formulated, anti-Semitism in Poland compared with that in Germany showed this difference: anti-Semitism in Poland came from those out of power, from the middle class, who were largely hostile to the government and who forced it on the government; anti-Semitism in Germany came from those in power, from the ruling class, and was forced upon the population. In Poland anti-Semitism penetrated upward, in Germany it penetrated downward. This is an oversimplification, but it is a fair first approximation.

The universities were strongholds of anti-Semitism. With the increasing poverty sons of workers and peasants lacked the means to attend the university. Thus the student body was composed largely of the middle class which hated Jews as unscrupulous competitors in commerce and offensive neighbors in the towns. The universities formed active centers of anti-Jewish and reactionary propaganda. The old traditional laws of the country, expressed in the slogans, "freedom of learning" and "autonomy of the universities," were used to fight democracy and to defend the strongholds of reaction.

My own case was an example of the difference between the attitudes of the government and the university. I had not had any difficulty in being acknowledged by the Ministry of Education as a docent. I should not have had any difficulty in being acknowledged by the Ministry as a professor. I was offered a position on a government paper. But my difficulties arose from the reactionary spirit of the universities. They had the right to appoint professors in the most democratic way, through faculty meetings, and skillfully turned the sword of democracy into a weapon of reaction.

Through my position on the *Gazeta Polska* I met some of the most powerful personalities in Poland, and I was less nervous than when meeting some of the professors of the university. The government was certainly anti-Semitic, but this anti-Semitism was rather a matter of policy than of Jew hatred, and it permitted individual exceptions; I was one of the exceptions. I was privileged. I earned money. I met the right people. And I felt the danger of satisfaction. I was tempted into an attitude into which many successful Jews have been tempted. I could say:

"Jews exaggerate the anti-Semitic issue enormously. I have the opportunity of lecturing at a Polish university and even have a position on a government paper. Perhaps it is a little more difficult for a Jew. But this is good, because it strengthens his character. The cry of anti-Semitism is often a cheap excuse for someone who is a failure and who would have been a failure anyhow."

It is very convenient reasoning. It helps to clear one's own conscience by ignoring a problem and denying its existence. It is the method often used by Jews who have become successful. I was very nearly affected by the same germ of security to the point of being smug and snobbish. Almost I pictured myself as a professor of a university whose lucrative hobby was writing, who had a peaceful life, quietly turning out two papers yearly, growing automatically in fat and in respectability.

A little over a year had passed since my return from England. The question of the professorship at Wilno was still undecided. I regarded my chances as fairly good. There were three steps prescribed by law: *one,* the dean must write to all professors of the same subject in Poland, asking whom they would recommend; *two,* the faculty selects one among these recommended candidates; *three,* the Ministry of Education approves or (very seldom) disapproves the choice. I knew that only the first stage of this involved procedure was completed.

During the Christmas vacation I went to Cracow to see my family. Before going I wrote a post card to W., the new professor of theoretical physics in Cracow, announcing my visit. We knew each other well. He was a type often met among scientists: very quick to grasp the essential point in discussion but too lazy for the persistent, continuous thought by which alone original results can be achieved. Isolated as he was from other physicists, Professor W. was glad of the opportunity of a scientific discussion and made an appointment with me in a café.

"What are you working on?" was his first question.

"Still on the unitary field theory. Recently I discovered that Maxwell's theory can be generalized in many different ways. This is not astonishing. But I no longer believe that the generalization presented by Born is the simplest. I don't like the

arbitrariness of the whole problem." I took a paper napkin and
began to write some formulae to explain my last results. Then
I changed the subject:

"By the way, how is Wilno? I hope you have already answered their question about a recommendation."

"Yes, I did," answered W. very quietly, absent-mindedly
looking at the formulae. "I recommended Professor C., your
superior, from Lwow."

"What?"

He looked up, holding the paper napkin in his hand, astonished at my violent reaction.

"I thought you knew. Didn't Doctor C. tell you? They decided that they wanted Professor C. from Lwow, and worked
upon him until he accepted even though it means less prestige for
him. They asked me to make it easy for them by recommending Doctor C. It is an old custom that, together with the official
letter asking for a recommendation, private letters are sent sug-
gesting a definite name for recommendation. As it was a purely
formal affair and I knew that my letter wouldn't influence the
issue, I thought I might as well help them."

He explained the incident as quietly and simply as though it
were the most natural thing in the world. He was astonished
that I had ever expected to become a professor in Wilno. He
regarded anti-Semitism as something which one must count on
and be prepared for, like bad weather. My strong reaction
seemed to him out of place. I ought to have learned better in the
thirty-eight years of my life.

University professor though he was, I could not help bursting
out with a speech which I still do not regret:

"You played a dirty trick on me, which was all the worse because you have pretended to be my friend. You know very well
that I am certainly no less fit for this position than Professor C.
But Wilno again decided to push C. above me, and it did not
occur to you to be loyal to me. You helped them to play the
trick and you regard your behavior as natural. You are even
astonished that I should dare to reproach you. In this small matter, in which you could have shown your decency and at least

passively opposed anti-Semitism, you accepted the anti-Semitic point of view, a view which sooner or later will turn against you. It begins with one hundred per cent Jews but it does not stop there. The next step is against the fifty per cent Jews like yourself. Acting as you did, you threw away the opportunity to show some decency. You understand that now there is no possibility of friendly relations between us."

Professor W. became red in the face.

"I guess you are right. But I did not see the problem in this way. It just did not enter my head."

Without awaiting further explanation I got up and left the table. Two days later I returned to Lwow. I had dinner with Professor C., and between the soup and the meat I remarked:

"I saw Professor W. in Cracow and learned some interesting gossip about Wilno."

Professor C. blushed, looked down at his napkin and mumbled:

"He told you, most probably, that I am going to Wilno."

"Yes. I decided to talk to you about it. I have some feeling of tidiness and should like to make my point of view clear, especially as our relations as colleagues here in Lwow have been correct. To stay or to leave Lwow is your personal affair in which I have no right to interfere, although I am a little surprised that you are willing to leave Lwow for an inferior university. But I want you to know that by accepting Wilno you kill my only chance of getting a professorship. I want you to know that I will have to wait twenty years before the chance is repeated, and then I will certainly miss it once more because I will be too old."

He did not raise his eyes from the table.

"I decided to go to Wilno because I have some relatives there."

I said: "I understand. This is an important reason, although I myself prefer to run away from places where I have relatives."

He felt how silly his answer sounded and added with a sudden outburst:

"They asked me to accept the professorship there. I recommended you, but they told me that they wouldn't take you."

"Why?"

I wanted to force him to give the simple answer: "Because you are a Jew." But instead I heard:

"Because they do not like your writing for the government paper *Gazeta Polska*."

The intrigue was carefully staged, much better than in the case of my docentship. First Professor C. was convinced that he ought to accept the invitation to Wilno to save the Polish character of the university. The rest was easy. C. would go to Wilno. His chair would then be free in Lwow. Professor Loria would have neither the courage nor the opportunity to push me in Lwow. There was then in Lwow a theoretical physicist, older than I, who was already a professor of the polytechnic school and who could simply be transferred to the university, as his chair of theoretical physics had recently been abolished by the Ministry of Education. The whole strategic plan was worked out in advance. It was like a series of moves in a chess game:

> Professor N. from Cracow retires.
> Professor W. from Wilno goes to Cracow.
> Professor C. from Lwow goes to Wilno.
> Professor R. from Lwow Polytechnic School goes to Lwow University.
> The chair of theoretical physics in Lwow Polytechnic School is liquidated.

No loopholes anywhere. No Jewish docents wanted. The plan was perfect and it demonstrated clearly that I was superfluous in Polish academic life.

Before making any decision I discussed the situation with Loria. He told me:

"You know how much I should like to help you because I know that you deserve it and that an injustice is being done to you. Anti-Semitism has increased tremendously of late. Our neighbor has set an example which people here would like to imitate and duplicate. The Jewish problem of three and a half million people is a hopeless one. They will be squeezed out economically if the social attitude doesn't change. The only hope is to belong to that thin layer for which exceptions are made. In this respect you are a privileged person. You have a good income from working here and writing, and you can easily afford to wait until the atmosphere changes and we shall be able to do something. But I cannot do anything at this moment. I don't believe in fighting for a hopeless cause."

"It is not so simple," I argued. "First I don't believe that it will be possible to retain my privileged position for any length of time. I could still be quite happy here with the docentship and my newspaper work. But anti-Semitism is like leprosy. Sooner or later it will affect the whole cultural life of our country if it is not checked by social change. The attack on the positions of Jews like me will come later, but it must come. And there is another argument against my waiting here hopelessly for a change, an argument more general and basic. I am beginning to be ashamed of my privileged position. The waves of hate are too strong. It does not help me to say that they did not splash me. I know that to go away is cowardice. But so it is to stay here and do nothing about it."

Loria disagreed with my last remark:

"You cannot do anything about the Jewish problem. In five hundred years historians will say 'After the great war the human race went through fifty years of darkness and barbarism.' This will perhaps be their only comment. But these are the fifty years of our lives, the only ones we have. It seems to me that the only solution is to do little decent, seemingly insignificant deeds. I was never considered much of a fighter. But by saving you from teaching in a gymnasium I did something good, something that was valuable. To be decent in the small field in which one is allowed to act means much and—believe me—it is quite unusual among university professors."

I felt that Loria wanted to keep me in Poland and I fought his arguments:

"I simply cannot bear this atmosphere of hate, of intrigue and discrimination any longer. You could say that I see it in such an exaggerated form now only because it has beaten me personally, that I should have closed my eyes to it had I been offered the silly chair in Wilno. You are right, and I despise myself for it. In some ways I am glad that I received this blow, that it saved me from self-satisfaction and snobbishness. I want to leave Poland. I cannot bear the feeling of being unwanted."

Loria became warmer, more personal:

"You are restless, and your tragedy is still responsible for it. I don't want to convince you, but you know how painful it will be for me to lose you; because I care about you personally and because I care about science in Poland. But I believe that the situation may change for the better as it recently changed for the worse. And then I should like to have you here."

I answered:

"I shall always be happy to come back whenever I feel that I am wanted. It is not easy trying to build up a new life in a world full of splendid scholars looking for jobs. You know all about the intrigues which preceded my docentship. The story has been repeated and will be repeated again in the future, as long as the political situation does not change. I don't know at this moment what to do, but I shall try to go away from here and not to give anyone the opportunity of reminding me that I am not wanted."

The political scene in Poland changed rapidly. One of the important factors which determined the general trend of events was Pilsudski's death. His personality had held together men with different social outlooks. In the year 1935–36, the last I spent in Poland, great changes began to occur. The friction within the government became common knowledge. The internal struggle was too bitter to be kept secret. I was in Warsaw for a conference with the editor of the *Gazeta Polska* on the day when an issue of the paper was confiscated by the government and later put out with the first page blank, the page on which the editorials were printed. Thus the government paradoxically suppressed its own paper. Thereupon Matuszewski, the editor, representing the liberal group, left *Gazeta Polska*. The opposition of the National Democrats grew as the administration became weaker. The anti-Semitic slogans of the opposition became louder with each day.

Workers demonstrated against the weak government. In Lwow the police shot into a workers' peaceful demonstration and the mob, making use of the chaos, smashed windows in Jewish houses and plundered Jewish shops. One day when I went from my peaceful street to the university through the center of town I came upon broken windows, plate-glass shop-windows smashed and replaced by wood, signs shattered by stones. Cafés were deserted and their great square windows broken. Lwow's gay streets were empty. Patrolling policemen and a few passers-by were the only visible signs of life on pavements covered with rubbish and broken glass. There was broken glass everywhere. It broke in still smaller pieces under the pressure of my shoes, and its crunching accompanied each step which led me through the empty streets. News of the riots circulated privately. Newspapers were forbidden to mention it.

The temperature at the universities rose to the boiling point. Even in our small department the tension and the anti-Semitic mood were apparent. Students belonging to the most brutal anti-Jewish societies wore small green badges on their jackets. In the beginning they were tactful enough not to wear them during my lectures. But on the streets I saw young girl students who had often come smiling to my office to ask for advice now wearing these green badges. Humble boys, studying mathematics or physics, proudly announced in green their attitude toward the Jewish problem.

Lectures were disturbed and Jews beaten by well-organized gangs. This was the usual procedure: a gang planning an attack on one department, physics for instance, was selected from a different school, such as the veterinarian. They would arrive suddenly, each carrying a stick. In the handles of some of the sticks a narrow groove had been cut and a razor blade inserted so that its sharp edge was barely visible. Having pushed the attendant away, they would rush to the lecture room and beat the Jewish students with the sticks and razor blades until the blood flowed. Then they would quickly vanish. The whole performance took a few seconds. All the details had been planned beforehand. In the lecture room there was always someone who would silently indicate the Jews to the gang leader.

From time to time deaths occurred. Jews organized in self-defense, and instances of the killing of non-Jewish students were reported. The professor was helpless, as the police were not allowed on the university campus. To call for the police would be to destroy the "freedom of learning" and the "autonomy of the university." The only thing a professor could do was to call for the "rector," representing the highest authority of the university. But by the time the rector arrived everything was over and he could only look at the wounds of the Jewish students, if they had not in the meantime been taken to the hospital.

The only weapon which the university senate used was the closing of the university. When the attacks began to be too frequent the rector suspended all lectures for several weeks in the hope that the youths would quiet down. The first week after the reopening everything was comparatively calm, but two or three weeks later the riots broke out again and lectures were again suspended. Teaching became a sad farce. I do not believe that in my last year in Poland I delivered even half of my scheduled lectures.

The big issue for which the students had fought for years was expressed in the slogan, "The University Ghetto." The reactionary student organizations demanded of the university authorities that special benches be set aside for Jews and that they be forbidden to sit near Gentiles. Before the university took its stand on this issue the students compelled the Jews to take seats on the left-hand side of the auditorium. The progressive Gentile students and Jews who ostentatiously attempted to sit together on the right-hand side were beaten and expelled by the better organized and physically stronger reactionary group. I was touched to hear that Pilsudski's daughter took her seat among the Jews.

One of the progressive professors, witnessing one of these disgraceful scenes before his lecture, said:

"If Jesus Christ came here, He would take his place among the oppressed."

After a long struggle the university authorities surrendered and designed special benches for Jews in all auditoria. The rector of Lwow University, a humane and liberal man, resigned over this issue. The Jews put up a hard fight, refusing to sit in "The University Ghetto." During the lectures they stood behind the benches and took notes, if not prevented by being thrown out. Having won this victory, the reactionary students pressed for more. They introduced Jewless days and then Jewless weeks and began to fight for *numerus nullus*, or no Jews at the university.

The government was carried away by the wave of anti-Semitism. By its own weakness it was forced to adopt anti-Semitism, whose purpose was twofold. On the one hand the rising anti-Semitic propaganda distracted the attention of the people from grave economic and social problems, from the misery of the peasants, from unemployment and poverty. This feature is common to all anti-Semitic waves rising everywhere. On the other hand the government attempted to weaken the power of the National Democrats by adopting their anti-Semitic slogans. Thus an inevitable situation arose. The Nationalist party sharpened its anti-Jewish slogans, claiming that the government had seen the light but not clearly enough. To which the government party then replied by sharpening theirs. A race began for the supremacy of those souls most possessed by hate, a race in which power was the prize and the destruction of Polish Jewry the only possible outcome.

Away—away from the air saturated by hate which darkened the sun and shadowed all my days! Away from the endless talks of the Jewish problem, from whispers of the still darker future and of lost hope. I did not feel heroic enough to stay and fight. I decided to try my luck in America. I had a little money put aside from my earnings on the *Gazeta Polska*. I was sure that my career as a contributor to this paper would end soon. My last article there appeared in the same issue in which Madagascar was suggested as a colony to which Polish Jewry should be transported. Although I, personally, had no difficulty either at the university or on the newspaper, I knew that it was only a matter of time, and I preferred to resign and go away rather than wait until I should be thrown out.

I wrote to Einstein, explaining my situation. I was well aware that it was one of the many appeals he received. A charming reply came, stating that the Institute for Advanced Study in Princeton had granted me a small fellowship and that he was looking forward to meeting me. In my thirty-eighth year I was to try again to build a new life in a new country.

XIV

DURING MY CHILDHOOD, and later in my youth, my thoughts and desires centered about a peaceful life, around teaching and research, around an occupation which promised uneventful, quiet days, in which emotions were drawn from studying, reading and contemplation.

But this life remained a mirage, a picture which faded the moment I felt that I might grasp it and keep it. I fought for a peace and rest which did not come, and I experienced a strange change, a substitution of means for ends. I thought that peace and security were my aims, but ambition and tension grew from my desires to achieve these ends. I thought that I was approaching my goal by a difficult and bitter road. Obstacles, defeat, victory and again defeat, all taught me to enjoy the fight with its strategy and its continual ups and downs. Through such a life the longing for peace may change into an aversion toward the boredom of harbors and havens, into a desire for new adventures.

Is this perhaps not the reason for the nervous restlessness characterizing Jews? Is this not an inevitable reaction against the obstacles and frustration which are so often their lot, so that they welcome contention and struggle?

To achieve peace of mind for scientific work I had to leave the country in which I was rooted. Perhaps I welcomed the defeat because I had developed a taste for a restless life. I burned my days hoping to accelerate the rhythm of time, to bring me nearer future events. I wanted each day to pass quickly in my impatience for tomorrow.

This was not my reaction alone. I developed the characteristics which grow in every member of a suppressed nation or race. Though, curiously enough, I do not believe that the same applies to a suppressed class.

An exploited worker who knows that he is exploited joins a union, fights for better conditions and a better social order. In his fight he finds an emotional outlet and hope for tomorrow. He identifies himself and his interests with those of his own class; he feels the support of the working class and knows that he is not alone. He may try the way up by leaving his factory, by educating himself, and he may succeed. If he does, he will not suffer for his worker's past. He always has in his background the class on which he can lean and before him the goal which he wants to achieve. If he despises the class from which he grew, he deserves contempt.

It is different for one who tries to escape from the ghetto. His driving force to escape is rooted in contempt for the environment from which he comes. I hated the Jewish school. I disliked the Jewish language. I was not attracted by the Jewish religion. These were the driving forces. I never had—as the worker has—the struggling background of the class from which he comes. The ghetto is full of misery, dirt, sadness, but there is not even so much as a spark of fight. Religion and obedience to the Almighty make the misery bearable and throw over life the weak luster of poetry.

From early childhood I nursed the same emotions toward my environment, and from those feelings I drew the strength which

carried me toward the outside world. The realization of this desire to escape formed the first serious problem from which I
learned to think and act, and it was this struggle for achievement
which formed my character.

To escape I was forced to fight against my environment. I
tried hard, and often cynically, to burn out of me the traces
left by my upbringing. Every successful step outside was bound
to increase my contempt for the small, sad world from which I
came and the desire to erase its visible signs. I did as a prisoner
does after escape. His first deed is to change his prisoner's garb
and rid himself of the chains still hanging about his ankles. As
brutal and ugly as this sounds, it is still better to acknowledge
the existence of this attitude than to be entangled in self-deceit.
Jews who succeed in leaving the ghetto must go through a
period in which they despise and detest their old world, because
these emotions form the driving force for escape. This contempt and the force directed outside are strongly connected:
they nourish and, by mutual induction, intensify each other.

Just before I left for America I went to Cracow. I wandered
through the ghetto of my town. On a summer morning the
voices of Jewish boys singing in chorus the words of the Torah
reached me through the open window of the school. "There
may be among them someone who hates this place as I hated
it and who dreams of going to a gymnasium." I went nearer.
The school windows were open, the first-floor windows of a
dreary house. I smelt the foul air of the room. It was the same
air, the same smell of onions and potatoes, which I had smelled
over thirty years before. I saw the tired, thin, badly nourished
faces with burning dark eyes, and for the first time in my life
I was conscious of a touch of poetry in this sad ghetto scene.

"What did I do to help them? I did nothing in the past and can
do little in the future. No one here will listen to my voice. We
no longer speak the same language. Most of them would spit in
my face and throw stones at me. But there may be one among
them whose eyes burn with desires, who will try the hard and
lonely path which seems to lead outside. If I meet him in life I

shall try to help him, because I know the hardships and sufferings of this long journey."

No! It is not true that I had only scorn and contempt for my environment. There is a curious mixture of hate and love. I could say to myself: "What do I care about Jews? I am above racial, religious problems and prejudices." But all my continued attempts to tear off the bonds only prove that these bonds exist, and they will exist to the last day of my life. Hate and scorn carry subtle overtones of love and attraction.

It is not easy to gain freedom and to escape. It is a bitter fight repeated over and over again, until one's system absorbs the necessity of fighting and the restlessness created by it destroys the peace so ardently desired but never to be attained. There is no harbor, no haven anywhere.

The struggle to leave the Jewish environment and to mix freely with non-Jews creates and must create the characteristic features for which Jews are despised. The bitter struggle deforms character. Anti-Semitism nourishes itself. It is a monster which increases and grows by the laws governing hate. Barriers created by anti-Semitism shape the characteristic features for which Jews are hated and by which anti-Semitism attempts to justify itself.

There are only two ways to solve the Jewish problem. One is the simple, brutal way—to destroy the suppressed. The other way is to destroy the suppression. The growing monster of anti-Semitism can be forced to kill itself; it can be nourished by reaction or annihilated by progress.

I know that my destiny is the destiny of thousands and that my feelings are reproduced in thousands of others. I felt that the roots of my life had been taken away with the earth of the country in which they grew. It is not true that I and others treated far more brutally are without a country. Poland is, and will remain, my country. I shall always long for the Polish fields and meadows, for the air smelling of flowers and hay, for vistas and sounds which can never be found elsewhere. I shall long for the countryside with its bad roads and peasants' huts, for Polish food and the smiles of Polish girls. And, above all, I shall long for the sound of the Polish language, the language in which I expressed my earliest joys and sorrows and incoherent silly thoughts—the only language in which I can make stupid puns, tell ribald stories and curse roundly. I will never forget my country. It is the co-ordinate system to which I refer the most important events of my life. There I spent the childhood which determined my future. There I met Halina and there I endured the day of her tragic death. There I was taught and there I wanted to teach. I shall try desperately to put out new roots in a new earth. My intentions may be strengthened by the friendliness and opportunities which will be offered me. But the mirage of my country, idealized by thought and vividly painted by the force of imagination, will always be before my eyes.

BOOK THREE

Search and Research

I

I REMEMBER A MOTIF which often appeared in the dreams of my childhood. It was the dream of a strange, remote country: America. It lay across one fantastic room. Within this room was a lake in which swans with long curved necks were swimming. The lake resembled a picture on the wall of a restaurant where I had once been taken by a rich uncle from New York. In my dreams I went to America to perform a special task and felt the heavy responsibility of my duties. It was disappointing to wake up to see my grandfather in one bed, my two sisters in another, and to realize that I should never see this far-off land. I thought of these dreams my last night on board ship. It was a sleepless night full of disorderly thoughts, of expectations and fears.

I was attracted to the United States, by its physical distance, by my desire to go away as far as possible. But now that I was farther than ever before from the scenes of my childhood, I felt closer than ever to my past.

"What do I know of the country to which I am sailing? I know that it is a materialistic, standardized country where talk centers around sports, clothes and the trivialities which lie on the surface of life. A country in which a European feels lonely and lost. I know this strange continent. The movies have brought to the smallest village in Poland a vivid picture of America. Hollywood shows so little imagination that the only thing which can be expected is realism. A train is a train, a house a house and a farm is a farm. It will be a difficult country for a

European to live in. I shall be deprived of all intellectual atmosphere; the conversation will be stiff and trivial. But it is perhaps better now than it was years ago. So many scholars have lately emigrated from the European universities and raised the standard of the funny American colleges where the most important thing is a childish game consisting of throwing a big leather egg. Universities which are private institutions are a strange invention anyhow. In Poland, Germany and France no one ever heard of anything like this.

Such was a brief summary of my thoughts. They were not exceptional. That is what Europeans usually think about America, bringing with their luggage feelings of superiority toward this young and undeveloped country whose intellectual level they have decided to raise by their presence.

It is also the land of advertisements, reiterated slogans and banal catchwords. The language is a spoiled English spoken through the nose, full of such phrases as: "Oh yeah," "Just too bad," "Says you," "Stick 'em up," "I guess," "Honey, I'm crazy about you."

All this I had learned from the movies. They did their best to confirm and strengthen the feeling of superiority of the old cultured Europe. We admitted we had no skyscrapers and refrigerators, but we believed we knew more about art, poetry, love-making, science and the intelligent enjoyment of leisure.

"But there is one good thing about the coarse, unsophisticated Americans. They seem to be a decent sort of people. They seem to be helpful. So many Europeans have come here lately and made a living. Shall I succeed? I don't want to go back to Poland. I am not wanted in my country. I am going to start over again, for the first time in my life not handicapped by being Jewish. There is no anti-Semitism in America. What a pleasure to breathe the air of a country without thinking that I am despised and not wanted."

I firmly believed that there was no anti-Semitism in America. I do not know why. I had heard statements about increasing anti-Semitism in the New World. But I flatly refused to believe them. Perhaps because of the dreams of my childhood, perhaps because of my conviction that there must be a country free of hatred and, if there was one, it could only be on a distant continent. This belief was an essential aspect of my American picture, compensating for all its unattractive features.

The only Americans I had met on the European continent happened to be bewildering creatures, full of self-satisfaction, running from one place to another, confusing countries and experiences and madly spending money.

I knew little of American social problems. I had heard about two political parties contending with each other though there was no difference in their programs; about the workers with unawakened social consciences, attracted by the middle-class standard of life which they tried to imitate. I had read about Roosevelt and the New Deal, which I did not even try to understand.

It was my last night on the ship. Full of excitement, I could not sleep. I dressed at five and went up on deck. I saw the land. There was nothing exotic in this view, which might have been anywhere in Europe. A few isolated houses, some trees, hardly visible in the gray light of the dawn.

I was waiting for the spectacle of the skyscrapers, seen so often in the movies. But nothing seems real until we can touch it with our eyes. The realistic photographs of skyscrapers seen in Polish movie houses appeared fantastic and unreal. They remained a shadowy dream until fate brought me physically near.

I had to arrange passport formalities, to explain that I was going to Princeton, to show the letter from Einstein inviting me to the Institute for Advanced Study. I had to wait a long time in the queue until my turn came. After finishing I hurried breathlessly to the deck, and there suddenly before my eyes, without any intermediate scenes, were the Statue of Liberty, the skyscrapers of Manhattan Island, the crowded ferries crossing the Hudson, the docks with their polyglot inscriptions. The view was strange, grotesque and magnificent. The sky line was a collection of blocks piled higher and higher by competing children. The tendency of America to formulate and solve *extremum* problems was clearly indicated in this view.

A short fat finger encircled by platinum and diamonds suddenly appeared in front of my nose, and I heard a shrill woman's voice: "There is the Empire State Building, the highest in the world. There is the Woolworth Building, which was the highest before the Empire State Building was built. There is——"

She talked like an auctioneer, offering to the highest bidder pieces of Manhattan's sky line. In return for your bid she would let you take the Empire State Building home in one pocket and all Rockefeller Center in the other.

A strange city standing erect. To look over the city one must move one's head vertically and not horizontally as in all other views. It is unique. The only other city which came to my mind was Venice, as different from New York as the times in which they were built. The strangeness of both views, the curious mixture of water and architecture, brought this comparison to my mind.

Not caring to explore the city then, I went straight to Pennsylvania Station. I passed busy, noisy, dirty sections of the town, saw in the distance shining skyscrapers.

There was a restlessness in the atmosphere. People rushed along energetically, hurrying noisily. How different it was from the way people moved and walked in London! Everything seemed to be built hurriedly and in a haphazard fashion, without any thought for the distant future, nervously erected to fill up empty spaces in the shortest time.

It was Saturday, about 1 P.M. I went by train to Princeton. Not knowing of the difference between coaches and Pullman, I felt deceived by the movies when I found myself in a shabby old day coach crowded with young men and girls. I was plunged into the noisy, nervous atmosphere which I thought to be so characteristic of America.

"Why are the trains so crowded? Is everyone going to Princeton for the week end? If so, where is their baggage? The week-end traffic should be in the opposite direction, from Princeton to New York. Why are all the people going to Princeton?"

I could find no explanation. The view through the windows

strengthened my impression of chaos, disorder and restlessness. I was astonished at so many wooden houses; in Europe they are looked down upon as cheap substitutes which do not, like brick, resist the attack of passing time. Old junked cars, piles of scrap iron, the speed with which we passed New York and reached the less densely populated areas, the provisional character of the buildings, all were so different from the peaceful, settled look of the English countryside. It was another world.

In my thoughts, for a long time, I had compared New York with London, Princeton with Cambridge. It never entered my head to compare anything with Poland. Was it because I wanted to suppress all thoughts of my country?

I arrived in Princeton. The crowd which had filled the train left it with me. The puzzle began to be more serious. I asked a man entering the station:

"Excuse me, could you tell me where is the mathematics department of the university?"

He looked at me, smiled pleasantly.

"I am sorry, I'm not a Princetonian."

I wanted to ask someone else, but in the few seconds of my conversation the crowd had vanished, and when I looked for someone else there was nothing but emptiness around me. Led by intuition, I found the campus. There were the university buildings separated by grass and trees which shone with the vivid colors of an American autumn. In a few hours I had been transported from the intense noise of the greatest city in the world to a place as quiet as a cemetery. No one, absolutely no one was on the campus. The silence was uncanny. Deserted buildings in a complete calm, and no one of whom to ask my way to the department of mathematics. And the buildings, grotesque at any time, looked still more strange when no one was around. For whom were they built? A few hours before I had marveled from the deck of the ship at the straight lines and modern architecture of some of the skyscrapers. Now I was seeing an amazing tutti-frutti of all possible and impossible styles. There were Cambridge and Oxford. There was an imitation of the Trinity College Gate. It was in stone instead of brick. The copy was supposed to be exact, but it looked grotesque whereas the original was beautiful and impressive. Was it the magic of the ages which had changed the color of the bricks, was it the influence of time which made the difference between the original and the imitation, shining with newness? Then I saw another building. It was in Venetian style on one side and Moroccan on another, with a balcony for Romeo and Juliet. Then two small shabby buildings in the style of Greek temples. The conglomeration reminded me of a European's idea of Hollywood: buildings of different styles lying about the movie sets. Even the physics building, for which the modern architecture of straight lines and large windows is best suited, was Gothic.

I wandered around the deserted campus looking for the department of mathematics, for Fine Hall, which is famous all over the world wherever mathematics and theoretical physics are taught. Just as I had given up hope of finding it in the maze of buildings, I saw an older man walking along the campus. This time I was lucky. A few minutes after I had obtained the information I stood before the place for which I had left Poland and crossed the Atlantic. It was a three-story Gothic building in red brick, simpler, much better than many others I had seen on the campus.

I imagine that there must exist some similarity between my emotions and those of a devout Catholic who enters a church after a long pilgrimage. Here is the center of mathematics of the world. Here Einstein, Bohr, Dirac and Schroedinger lectured. It would be easier to name those of the great mathematicians and theoretical physicists who have not lectured in Fine Hall.

The door was open; I entered the corridor and read the names of the professors and the numbers of their rooms. What a splendid collection! Among them was: Einstein, Number 209.

But Fine Hall was empty. No one was in. I went through the corridors, found the offices closed and the common room without a living soul. I had left New York hurriedly, to be in Princeton as soon as possible, only to find names, walls and a cemetery of stone. I wanted to leave. The uncanny atmosphere had begun to get on my nerves. The change from the noise of

New York to the deathly quiet of Princeton was too much. I passed Nassau Hall, the central building of the university, the only building in a pleasant, quiet, colonial style; the only building which does not pretend to be anything but simple, serious and sincere.

From Nassau Hall I could see the town. A weak spark of life remained there—the shops were open, a few people walked through the streets, a car passed—but there was still an almost painful air of quiet and expectation. I looked at a room in a hotel. The hotel might just as well have stood in a Polish provincial town; it was old and primitive. A little old lady led me through the empty hotel in which she and I and a colored porter seemed to be the only living souls. I could not bear the mystery any longer and I asked the landlady:

"Where are the students? Are they all gone?"

"Oh yes—all of them."

"May I ask you where they went?"

"I really could not say. Perhaps to see Notre Dame."

Was I crazy? Notre Dame is in Paris. Here is Princeton with empty streets. What does it all mean? Then suddenly: "What an idiot I am! If I show the same power of deduction during the year as I have shown today, I had better go back!"

Suddenly the whole atmosphere changed. It happened in a discontinuous way, in a split second. Cars began to run, crowds of people streamed through the streets, noisy students shouted and sang, boys loudly hawked newspapers. The quietness changed into wild excitement. Far into the night it was still impossible to sleep. Songs, laughter, ribald voices echoed through the town until morning.

This was my first contact with Princeton. On my first day I discovered the importance of football in an American college.

II

Once upon a time the most famous center of mathematics was in Goettingen, Germany. To study in Goettingen, or at least to attend for a short time, was as essential for a mathematician as for a Mohammedan to visit Mecca. The small town breathed mathematics. Shopkeepers did not press a student for money if he proved an interesting mathematical theorem. Together with its splendid mathematical atmosphere Goettingen cultivated thousands of stories which circulated among mathematicians and a tremendous snobbery.

The fame of Goettingen was associated in the past with the name of Gauss, one of the greatest mathematicians of all ages, with the versatile mathematician Klein, with the great Hilbert, now retired, with Weyl who is at present in Princeton, with the splendid woman mathematician Emmy Noether who, expelled from Germany, died in America, and with a crowd of docents who habilitated and taught in Goettingen, later receiving professorships in many other universities.

Now Goettingen is dead; it took a hundred years to build it and one brutal year to destroy it. The most famous school of mathematics in the world is now in Princeton, too new to develop a tradition but old enough to create a legend and its own peculiar snobbery.

Fine Hall is the name of the building which houses the mathematics department of the university. It is named in memory of the former dean and head of the mathematics department who died in an accident some years ago. One of his classmates gave enough money to erect this splendid building.

Fine Hall housed (until 1939) two schools of mathematics. There was the mathematics department of Princeton University and the school of mathematics of the Institute for Advanced

Study. The department of mathematics of the university is simply a very good department of a rich, distinguished university which had luck in building up this particular department. It is undoubtedly one of the best, but not unique in this country. What made Princeton so spectacular was just the peculiar mixture between the university and the Institute for Advanced Study, the combination in one building of two great schools of mathematics, creating an island in the world with the greatest density of mathematical brains. (Since 1939 the Institute for Advanced Study has its own separate building in Princeton.)

The Institute for Advanced Study has a history worth recording, although I am not sure that the details I am writing down here are true. The story circulating among mathematicians all over the world is perhaps more interesting than the real story which I am unable to tell. I heard it for the first time in Cambridge soon after the institute was created.

Dr Abraham Flexner was for a long time a teacher in a high school. When he was forty he started a new life. He went to Europe to study the organization of German and English universities. After having written his doctor's thesis on this subject he was entrusted by the Rockefeller Foundation with the study of the organization of American universities. His book, *Universities, American, English, German*, appeared in 1930, published by the Oxford University Press. The book contained a smashing criticism of some of the American universities, of their commercialism and betrayal of the spirit of learning, of theses written on ridiculous subjects. Here are two of the many examples from his book:

THE UNIVERSITY OF CHICAGO
A TIME AND MOTION COMPARISON ON FOUR
METHODS OF DISHWASHING

A DISSERTATION
SUBMITTED TO THE GRADUATE FACULTY
IN CANDIDACY FOR THE DEGREE OF
MASTER OF ARTS

Nor does the resemblance to scientific research cease with the title page. The dissertation includes: Introduction, Review of Literature, Purpose, Limitations, Method of Procedure, Results and Comparisons, Conclusions and Recommendations, Conclusions (once more!), Bibliography. Time was kept and motions counted for "the removal of dishes from the table to tea cart" and similar operations. In the washing of dishes, motions were counted and tabulated for "approach stove, grasp teakettle, remove lid at stove," "travel to sink, turn hot-water faucet on," "turn hot-water faucet partially off," "travel to stove, replace lid, turn fuel on," "approach sink," etc., over a total of a hundred typewritten pages. . . .

and:

Columbia's showing does not differ in kind. Through Teachers College, Columbia offers courses in "methods of experimental and comparative cookery," . . . "tearoom cookery," in "food etiquette and hospitality," in the "principles of home laundering." . . . "Abstracts on Recent Research in Cookery and Allied Subjects," prepared by an instructor but introduced by a professor, deals in Part I with "refrigeration." I quote the following:
"The theory of refrigeration implies the reduction of the temperature of a body below that of the surrounding environment."
"Frequent opening and closing of the doors of the food chamber causes an increase in ice consumption and a temporary rise in the temperature of the food chamber."
"A good refrigerator should have a temperature satisfactory for the preservation of food."
In Part II of the same series, "Research Abstracts" devoted to "ice cream," an expectant public is informed that, as respects "the influence of sugar," the primary function of sugar in ice cream is to sweeten it. . . .

According to the legend, which probably is not true, some time after the book appeared Flexner was called to the telephone by someone who, without giving his name, said:
"I was very much impressed by your book about universities. Do you think that you could build something much better if you had the chance?"
"I believe I could try, but why does it concern you?"
"Because I should like to give you a few millions to build a new educational institute."
This conversation is supposed to be the origin of the institute. If the story is not true it still represents the European view of

American philanthropy and of the unexpected character of events on this continent.

On the first page of the yearly bulletin of the institute the following is printed:

Extract from the letter addressed by the Founders to their Trustees, dated Newark, New Jersey, June 6, 1930.

"It is fundamental in our purpose, and our express desire, that in the appointments to the staff and faculty, as well as in the admission of workers and students, no account shall be taken, directly or indirectly, of race, religion or sex. We feel strongly that the spirit characteristic of America at its noblest, above all, the pursuit of higher learning, cannot admit of any conditions as to personnel other than those designed to promote the objects for which this institution is established, and particularly with no regard whatever to accidents of race, creed or sex."

The institute started with a school of mathematics on October 2, 1933. Later a school of economics and politics, a school of humanistic studies were added but they are less famous, less developed than that of mathematics. This raises an interesting question: why did the institute start with a school of mathematics although Flexner is not a mathematician? Here again the legend offers an explanation.

Working in the true American spirit in which everything that is good must be the best, in which only solutions of *extremum* problems count, Flexner wanted to get the best men for the institute. He inquired among mathematicians: who are the best mathematicians in the world? He inquired among philologists: who are the best philologists in the world? And among sociologists: who are the best sociologists in the world? But from the sociologists and philologists he obtained confusing answers, as many different lists of great names as there were people he asked. Only in mathematics was the situation different. The lists repeated the same names over and over again. If the story is not true it had to be invented to show that agreement in mathematics is much easier than anywhere else.

Although there are splendid names in mathematics in the Institute for Advanced Study, famous names known to every mathematician, they mean very little to the outside world. There is only one name which spread the fame and glory of the Institute for Advanced Study to the outside world: that of Einstein.

The institute forms a unique experiment, but it is difficult to say how it will work in the long run. A few of its chairs are occupied by very famous mathematicians whose official duties are nil. They may or may not lecture. Usually they do. They are completely free to do whatever they like. It sounds wonderful: the ideal life for a scientist. The struggle for daily bread need not bother the well-paid institute professor. He can use his whole time for research. The arrangement seems to be based on the assumption: the better the external circumstances, the greater the scientific achievement. But is it really so? Everything possible was done to form in Fine Hall an isolated system and to remove from it the impact of life and external struggles. Are these the best conditions for increasing scientific achievement? Is it not the very impact of the external events, the fact that one is in the middle of an active world, that one takes part in the great play of impinging forces which keeps the imagination active, thought vivid and prevents sterility? Like all the questions "What would have happened if . . .?" this question is perhaps meaningless. But the problem of how to achieve the highest possibilities is, I believe, not so simple. Isolation, comfort and security as a reward for work done in the past may destroy the circumstances in which and through which this work was done. Scientific achievement is, as Einstein so often remarked to me, a matter of character, and character is formed and developed by the hard struggles of life. Isolation, security, ivory towers, all may prove just as dangerous, or even more so, than too much hardship and bitter fights which destroy the conditions of work.

Here in the institute a most interesting experiment is being performed. A few of the most distinguished scientists are put in the best possible position. It is a social experiment on a small scale. It is certainly too early to judge the outcome. But even the basic idea of the experiment may be open to serious doubt. It will be interesting to watch the results and to note in the years to come the Institute's contribution to our cultural life.

The very few permanent professors of the institute are free of duties. Most of them lecture two or three hours weekly on very advanced subjects connected with their scientific research. Einstein does not lecture regularly but spends some of the morning hours in his room at Fine Hall, ready to discuss problems with any one of the institute who wishes to do so.

Besides the few permanent members there are temporary ones who come and go. The first year I was in Princeton there were about fifty of them in the school of mathematics. Anyone who has done good scientific work may become a member of the institute. It is something of a supergraduate school. Temporary appointments are given to scientists from all countries. Some of them, invited for a year of lectures, are the most distinguished professors in the world, for example: Dirac, Bohr, Schroedinger, Pauli. Some are promising younger scientists, from America or Europe, for whom the institute provides temporary fellowships. Some of them are university professors who spend their leaves of absence in Princeton. And finally some of them are political refugees from Germany, supported for a year or two by the institute until they find more permanent positions in American colleges. It is a heterogeneous crowd.

For the first year I obtained a small fellowship, and it was difficult for me to classify myself as a member of a definite subgroup. I did not belong to the subgroup of refugees for whom a retreat to their own country was closed. Actually I was a docent of the University of Lwow, on leave of absence, and could still go back. I did not belong to the subgroup of distinguished professors because I was neither distinguished nor a professor. Nor did I belong to the group of brilliant young men.

Still with an inferiority complex earned by my past and by my last defeat in Poland, I entered Fine Hall, the best place in the world to intensify this complex. My first impression was that this was the most wonderful place for scientific work. If you cannot do research here, then you cannot do it at all. Nowhere else would you have the chance to meet so many people who understand your problems. On whatever you happen to work there is a great probability of finding someone who is a better specialist in this particular field than you are. Your time is your own. There are no duties here. But you are expected to do some research and you are expected to produce a paper in a scientific journal. Nobody forces you to do so; it is not your formal obligation. But you are well aware that if nothing is published by you in the next year or two the fact will not remain unnoticed. This attitude is nothing new and not peculiar to Fine Hall. It is the same everywhere. The university expects! Your head of department expects! Your colleagues expect! But the pressure is much stronger in America than in Europe. It may be illustrated by this authentic fragment of an interview between a department head and a famous scholar looking for a job:

The head of department asks:

"What is your average production?"

The candidate looks puzzled. He does not understand the question.

"It is very simple. How many papers have you published?"

"I never counted them. But I believe somewhere around twenty-four."

"And how long have you been in the teaching profession?"

"Seven years."

"Now let us see. Twenty-four divided by seven is three point four. We can say that your rate of production is between three and four papers per year."

"I see."

"Now are you sure that you will be able to keep up this rate of production in the future?"

I might add that the candidate obtained the appointment although he refused to guarantee the rate of his intellectual production.

In spite of the pressure there is no reason to be afraid. If you don't do scientific work you will very likely hear no criticism from your colleagues. They will carefully omit the subject in their conversations. They will gradually cease to discuss scientific matters with you in order not to hurt you. They will be so noble and subtle, skirting the subject, that you will wish you had never come to this place. You do not have to learn this in Princeton. You know it before even entering Fine Hall, you know it from your previous experience.

I thought: "I must do some scientific work here. Shall I succeed? Shall I have ideas? And if I have them, will anything come of them?"

I saw myself working hard, looking desperately for fresh inspiration in this tense atmosphere. Every piece of research is partially a gamble. At the end of a year you may find that your idea was wrong or that someone else has done it in the meantime, though a little differently. You may try to boast of your efforts and explain your bad luck; your colleagues will listen politely and tell you about their own similar experiences, but you never will be quite sure that they really believe you.

This nervous search for results is known to everyone who has had to change environment and must prove again his ability for research. Every measure creates countermeasures. There are some ways of playing safe. One of my colleagues, a splendid scientist, on leaving Poland for a Rockefeller fellowship, told me in secret:

"I have just obtained an interesting result but I won't publish it. I shall keep it for an emergency in case I do not produce any results during the time of my fellowship. It may save me from disgrace."

My own position was peculiar. During the three preceding years, first in England with Born and later in Poland alone, I had worked on the unitary field theory. I was convinced that the subject was, essentially, exhausted and did not believe that it was worth while to pursue the theory further. Especially in Princeton there was rather a feeling against it. Einstein had noticed some profound difficulties in the unitary field theory; many others were quite skeptical, although practically everyone knew the papers, which means something in these days of narrow specialization. I intended to work on other problems. I had no paper in my drawer and I did not have definite ideas for future research. The normal strain of every scientist entering a new institution was amplified in my case, as I felt I had exhausted my plans for original work.

There was still another factor which increased my strain: "I am in America and I should like to remain here. I have to find a job. I have not the slightest idea what to do and (what is more important) what not to do about it. I don't know anyone here who can help me. I am not distinguished enough for the universities to run after me and not young and inexperienced enough to accept any little job. The eight years I spent in the Jewish gymnasia were practically lost to scientific activity. These are *the* eight years in which imagination is keenest, in which scientists usually do their best work. I should not have lost them if I had not been a Jew. Here I am, at the age of thirty-eight, in a new country, having to start all over again, to struggle again for tomorrow. I must do this because I am a Jew and could not find opportunities in Poland. And for the same reason I was deprived of a great part of the scientific capital which I now need to build a new life. I was deprived of earning my scientific reputation at the age when normally one builds his most solid foundations for the future. It is not a matter of eight years' difference. It is a matter of *the* eight years."

I did not think of all this consciously when entering Fine Hall. I knew it all the time. These thoughts were always at the back of my mind, emerging painfully in spite of my constant attempts to suppress them.

III

I CAME TO PRINCETON on a Saturday, lived through a dead Sunday and entered the office of Fine Hall on Monday, to make my first acquaintances. I asked the secretary when I could see Einstein. She telephoned him, and the answer was:

"Professor Einstein wants to see you right away."

I knocked at the door of 209 and heard a loud "*herein.*" When I opened the door I saw a hand stretched out energetically. It was Einstein, looking older than when I had met him in Berlin, older than the elapsed sixteen years should have made him. His long hair was gray, his face tired and yellow, but he had the same radiant deep eyes. He wore the brown leather jacket in which he has appeared in so many pictures. (Someone had given it to him to wear when sailing, and he had liked it so well that he dressed in it every day.) His shirt was without a collar, his brown trousers creased, and he wore shoes without socks. I expected a brief private conversation, questions about my crossing, Europe, Born, etc. Nothing of the kind:

"Do you speak German?"

"Yes," I answered.

"Perhaps I can tell you on what I am working."

Quietly he took a piece of chalk, went to the blackboard and started to deliver a perfect lecture. The calmness with which Einstein spoke was striking. There was nothing of the restlessness of a scientist who, explaining the problems with which he has lived for years, assumes that they are equally familiar to the listener and proceeds quickly with his exposition. Before going into details Einstein sketched the philosophical background for the problems on which he was working. Walking slowly and with dignity around the room, going to the blackboard from time to time to write down mathematical equations, keeping a dead pipe in his mouth, he formed his sentences perfectly. Everything that he said could have been printed as he said it and every sentence would make perfect sense. The exposition was simple, profound and clear.

I listened carefully and understood everything. The ideas behind Einstein's papers are aways so straightforward and fundamental that I believe I shall be able to express some of them in simple language.

There are two fundamental concepts in the development of physics: *field* and *matter.* The old physics which developed from Galileo and Newton, up to the middle of the nineteenth century, is a physics of matter. The old mechanical point of view is based upon the belief that we can explain all phenomena in nature by assuming particles and simple forces acting among them. In mechanics, while investigating the motion of the planets around the sun, we have the most triumphant model of the old view. Sun and planets are treated as particles, with the forces among them depending only upon their relative distances. The forces decrease if the distances increase. This is a typical model which the mechanist would like to apply, with some unessential changes, to the description of all physical phenomena.

A container with gas is, for the physicist, a conglomeration of small particles in haphazard motion. Here—from the planetary system to a gas—we pass in one great step from "macrophysics" to "microphysics," from phenomena accessible to our immediate observation to phenomena described by pictures of particles with masses so small that they lie beyond any possibility of direct measurement. It is our "spiritual" picture of gas, to which there is no immediate access for our senses, a microphysical picture which we are forced to form in order to understand experience.

Again this picture is of a mechanical nature. The forces among the particles of a gas depend only upon distances. In the motions of the stars, planets, gas particles, the human mind of the nineteenth century saw the manifestation of the same mechanical view. It understood the world of sensual impressions by forming pictures of particles and assuming simple forces acting among them. The philosophy of nature from the beginning of physics to the nineteenth century is based upon the belief that to understand phenomena means to use in their explanation the concepts of particles and forces which depend only upon distances.

To understand means always to reduce the complicated to the simple and familiar. For the physicists of the nineteenth century, to explain meant to form a mechanical picture from which the phenomena could be deduced. The physicists of the past century believed that it is possible to form a mechanical picture of the universe, that the whole universe is in this sense a great and complicated mechanical system.

Through slow, painful struggle and progress the mechanical view broke down. It became apparent that the simple concepts of particles and forces are not sufficient to explain all phenomena of nature. As so often happens in physics, in the time of need and doubt, a great new idea was born: that of the *field*. The old theory states: particles and the forces between them are the basic concepts. The new theory states: changes in space, spreading in time through all of space, are the basic concepts of our descriptions. These basic changes characterize the *field*.

Electrical phenomena were the birthplace of the field concept. The very words used in talking about radio waves—*sent, spread, received*—imply changes in space and therefore *field*. Not particles in certain points of space, but the whole continuous space forms the scene of events which change with time.

The transition from particle physics to field physics is undoubtedly one of the greatest, and, as Einstein believes, *the* greatest step accomplished in the history of human thought. Great courage and imagination were needed to shift the responsibility for physical phenomena from particles into the previously empty space and to formulate mathematical equations describing the changes in space and time. This great change in the history of physics proved extremely fruitful in the theory of electricity and magnetism. In fact this change is mostly responsible for the great technical development in modern times.

We now know for sure that the old mechanical concepts are insufficient for the description of physical phenomena. But are the field concepts sufficient? Perhaps there is a still more primitive question: I see an object; how can I understand its existence? From the point of view of a mechanical theory the answer would be obvious: the object consists of small particles held together by forces. But we can look upon an object as upon a portion of space where the field is very intense or, as we say, where the energy is especially dense. The mechanist says: here is the object localized at this point of space. The field physicist says: field is everywhere, but it diminishes outside this portion so rapidly that my senses are aware of it only in this particular portion of space.

Basically, three views are possible:

1. The mechanistic: to reduce everything to particles and forces acting among them, depending only on distances.

2. The field view: to reduce everything to field concepts concerning continuous changes in time and space.

3. The dualistic view: to assume the existence of both matter and field.

For the present these three cases exhaust the possibilities of a philosophical approach to basic physical problems. The past generation believed in the first possibility. None of the present generation of physicists believes in it any more. Nearly all physicists accept, for the present, the third view, assuming the existence of both matter and field.

But the feeling of beauty and simplicity is essential to all scientific creation and forms the vista of future theories; where does the development of science lead? Is not the mixture of field and matter something temporary, accepted only out of necessity because we have not yet succeeded in forming a consistent picture based on the field concepts alone? Is it possible to form a pure field theory and to create what appears as matter out of the field?

These are the basic problems, and Einstein is and always has been interested in basic problems. He said to me once:

"I am really more of a philosopher than a physicist."

There is nothing strange in this remark. Every physicist is a philosopher as well, although it is possible to be a good experimentalist and a bad philosopher. But if one takes physics seriously, one can hardly avoid coming in contact with the fundamental philosophic questions.

General relativity theory (so called in contrast to special relativity theory, developed earlier by Einstein) attacks the problem of gravitation for the first time since Newton. Newton's theory of gravitation fits the old mechanical view perfectly. We could say more. It was the success of Newton's theory that caused the mechanical view to spread over all of physics. But with the triumphs of the field theory of physics a new task appeared: to fit the gravitational problem into the new field frame.

This is the work which was done by Einstein. Formulating the equations for the gravitational field, he did for gravitational theory what Faraday and Maxwell did for the theory of electricity. This is of course only one aspect of the theory of relativity and perhaps not the most important one, but it is a part of the principal problems on which Einstein has worked for the last few years and on which he is still working.

Einstein finished his introductory remarks and told me why he did not like the way the problem of a unitary field theory had been attacked by Born and me. Then he told me of his unsuccessful attempts to understand matter as a concentration of the field, then about his theory of "bridges" and the difficulties which he and his collaborator had encountered while developing that theory during a whole year of tedious work.

At this moment a knock at the door interrupted our conversation. A very small, thin man of about sixty entered, smiling and gesticulating, apologizing vividly with his hands, undecided in what language to speak. It was Levi-Civita, the famous Italian mathematician, at that time a professor in Rome and invited to Princeton for half a year. This small, frail man had refused some years before to swear the fascist oath designed for university professors in Italy.

Einstein had known Levi-Civita for a long time. But the form in which he greeted his old friend for the first time in Princeton was very similar to the way he had greeted me. By gestures rather than words Levi-Civita indicated that he did not want to disturb us, showing with both his hands at the door that he could go away. To emphasize the idea he bent his small body in this direction.

It was my turn to protest:

"I can easily go away and come some other time."

Then Einstein protested:

"No. We can all talk together. I shall repeat briefly what I said to Infeld just now. We did not go very far. And then we can discuss the later part."

We all agreed readily, and Einstein began to repeat his introductory remarks more briefly. This time "English" was chosen as the language of our conversation. Since I had heard the first part before, I did not need to be very attentive and could enjoy the show. I could not help laughing. Einstein's English was very simple, containing about three hundred words pronounced in a peculiar way. He had picked it up without having learned the language formally. But every word was understandable because of his quietness, slow tempo and the distinct, attractive sound of his voice. Levi-Civita's English was much worse, and the sense of his words melted in the Italian pronunciation and vivid gestures. Understanding was possible between us only because mathematicians hardly need words to understand each other. They have their symbols and a few technical terms which are recognizable even when deformed.

I watched the calm, impressive Einstein and the small, thin, broadly gesticulating Levi-Civita as they pointed out formulae on the blackboard and talked in a language which they thought to be English. The picture they made, and the sight of Einstein pulling up his baggy trousers every few seconds, was a scene, impressive and at the same time comic, which I shall never forget. I tried to restrain myself from laughing by saying to myself:

"Here you are talking and discussing physics with the most famous scientist in the world and you want to laugh because he does not wear suspenders!" The persuasion worked and I managed to control myself just as Einstein began to talk about his latest, still unpublished paper concerning the work done during the preceding year with his assistant Rosen.

It was on the problem of gravitational waves. Again I believe that, in spite of the highly technical, mathematical character of this work, it is possible to explain the basic ideas in simple words.

The existence of electromagnetic waves, for example, light waves, X rays or wireless waves, can be explained by one theory embracing all these and many other phenomena: by Maxwell's equations governing the electromagnetic field. The prediction that electromagnetic waves *must* exist was prior to Hertz's experiment showing that the waves *do* exist.

General relativity is a field theory and, roughly speaking, it does for the problem of gravitation what Maxwell's theory did for the problem of electromagnetic phenomena. It is therefore apparent that the existence of gravitational waves can be deduced from general relativity just as the existence of electromagnetic waves can be deduced from Maxwell's theory. Every physicist who has ever studied the theory of relativity is convinced on this point. In their motion the stars send out gravitational waves, spreading in time through space, just as oscillating electrons send out electromagnetic waves. It is a common feature of all field theories that the influence of one object on another, of one electron or star on another electron or star, spreads through space with a great but finite velocity in the form of waves. A superficial mathematical investigation of the structure of gravitational equations showed the existence of gravitational waves, and it was always believed that a more thorough examination could only confirm this result, giving some finer features of the gravitational waves. No one cared about a deeper investigation of this subject because in nature gravitational waves, or gravitational radiation, seem to play a very small role. It is different in Maxwell's theory, where the electromagnetic radiation is essential to the description of natural phenomena.

So everyone believed in gravitational waves. In the previous two years Einstein had begun to doubt their existence. If we investigate the problem superficially, they seem to exist. But Einstein claimed that a deeper analysis flatly contradicts the previous statement. This result, if true, would be of a fundamental nature. It would reveal something which would astound every physicist: that field theory and the existence of waves are not as closely connected as previously thought. It would show us once more that the first intuition may be wrong, that deeper mathematical analysis may give us new and unexpected results quite different from those foreseen when only scratching the surface of gravitational equations.

I was very much interested in this result, though somewhat skeptical. During my scientific career I had learned that you may admire someone and regard him as the greatest scientist in the world but you must trust your own brain still more. Scientific creation would become sterile if results were authoritatively or dogmatically accepted. Everyone has his own intuition. Everyone has his fairly rigidly determined level of achievement and is capable only of small up-and-down oscillations around it. To know this level, to know one's place in the scientific world, is essential. It is good to be master in the restricted world of your own possibilities and to outgrow the habit of accepting results before they have been thoroughly tested by your mind.

Both Levi-Civita and I were impressed by the conclusion regarding the nonexistence of gravitational waves, although there was no time to develop the technical methods which led to this conclusion. Levi-Civita indicated that he had a luncheon appointment by gestures so vivid that they made me feel hungry. Einstein asked me to accompany him home, where he would give me the manuscript of his paper. On the way we talked physics. This overdose of science began to weary me and I had difficulty in following him. Einstein talked on a subject to which we returned in our conversations many times later. He explained why he did not find the modern quantum mechanics aesthetically satisfactory and why he believed in its provisional character which would be changed fundamentally by future development.

He took me to his study with its great window overlooking the bright autumn colors of his lovely garden, and his first and only remark which did not concern physics was:

"There is a beautiful view from this window."

Excited and happy, I went home with the manuscript of Einstein's paper. I felt the anticipation of intense emotions which always accompany scientific work: the sleepless nights in which imagination is most vivid and the controlling criticism weakest, the ecstasy of seeing the light, the despair when a long and tedious road leads nowhere; the attractive mixture of happiness and unhappiness. All this was before me, raised to the highest level because I was working in the best place in the world.

IV

I was very much impressed by the ingenuity of Einstein's most recent paper. It was an intricate, most skillfully arranged chain of reasoning, leading to the conclusion that gravitational waves do not exist. If true, the result would be of great importance to relativity theory.

It is not easy to check someone else's reasoning. Only one who has done research knows the traps into which the greatest minds can fall. Bohr once remarked that an expert is a man who by bitter experience has learned of all the possible mistakes in his restricted field. To think, for example, that Einstein could never be wrong means not to understand what scientific work means. The greatness of Einstein lies in his tremendous imagination, in the unbelievable obstinacy with which he pursues his problems. Originality is the most essential factor in important scientific work. It is intuition which leads to unexplored regions, intuition as difficult to explain rationally as that by which the oil diviner locates the wealth hidden in the depths of the earth.

There is no great scientific achievement without wandering through the darkness of error. The more the imagination is restricted, the more a piece of work moves along a definite track —a process made up rather of additions than essentially new ideas —the safer the ground and the smaller the probability of error. There are no great achievements without error and no great man was always correct. This is well known to every scientist. Einstein's paper might be wrong and Einstein still be the greatest scientist of our generation.

On the same day that I talked to Einstein I went to tea in Fine Hall. There I met H. P. Robertson, whose work on the theory of relativity, especially cosmology, I knew well. He was a professor of theoretical physics at the university. His plump face would be expressionless if not for the clever and ironic eyes.

I talked with Robertson about science, and from the start we understood one another perfectly. I told him about Einstein's new paper, that I had not finished it but that the result still seemed strange to me.

"No," said Robertson, "I don't believe in the result. There must be a mistake somewhere. The gravitational waves exist all right, I am sure."

We continued our discussion for a long time in Robertson's office; it was pleasant to detect the frequent harmony of our ideas in physics and the agreement of our judgment. Later we shifted to more personal subjects. Robertson asked:

"Did you know Loria in Poland?"

I felt like answering "And how!" but instead said:

"Yes! He is my very good friend. Where did you meet him?"

"I took my Ph.D. in Pasadena when he was there. I liked him very much."

"Among other favors, I owe it to Loria that I went to Cambridge, where I worked with Born."

Robertson opened his wonderfully organized filing cabinet, took out notes concerning the unitary field theory and sat down comfortably, smoking a cigar and putting his legs on the desk.

"I read your papers with Born." He showed me the notes he had made of his critical remarks and went on:

"I know Born from my year in Goettingen. How did you get along with him?"

"We began with a quarrel. But then the next time we met he was angelic, and we lived happily ever after."

Robertson's experience was a little different.

"I started with a quarrel too. At that time I had my first ideas about the expanding universe. He thought that it was rubbish, and I did not care to see him again. I did not give him a chance to be angelic."

"You went to Goettingen to work with Born and gave him up after one quarrel?"

"Yep. I worked for myself, and I had a very good time too."

"Oh yes, I remember. Weren't you the man about whom Born told me a story? You went to a shop and instead of asking: '*Fräulein haben sie eine Wage, ich möchte etwas wiegen?*' you asked '*Fräulein haben sie eine Wiege ich möchte etwas wagen?*' "

"Rubbish. I made up the story and they believed it in Goettingen."

Then our conversation shifted again toward astrophysics and the astronomer Milne, upon whom Robertson delivered a long polemic and whose cosmological papers he criticized severely. It was refreshing to meet someone with a vivid, orderly mind, who enjoyed spiteful gossip and with whom talking was easy and pleasant. I hoped that I should see a lot of him. In this one building in one day, I met three mathematical physicists, one of them known as the creator of the theory of relativity, two of them great experts famous for their valuable contributions to this theory: Einstein, Levi-Civita, Robertson. What a wonderful place to work! I knew after my experience in Leipzig and Cambridge how essential it was for me to be in a stimulating atmosphere and to have the opportunity of talks and discussions. This time I had everything that I could desire.

I studied Einstein's complicated paper carefully. I thought that if Einstein's conclusion were true then there must be a simpler way of showing it. I tried, as every scientist does, to transpose the problem into my own way of thinking. After a few days of work I believed I had a much simpler proof of Einstein's result that gravitational waves do not exist. Such a simplification as I thought I had made is a neat but rather small matter. The method of proof is not so essential as the fact that an important theorem is true. To formulate a theorem a vivid imagination may be needed. Usually the first proof is done in a roundabout way. The simplification of a proof is something much less valuable. It is rather secondhand—important, perhaps, from a didactic point of view— but it is on a different plane from the formulation of an interesting theorem requiring creative intuition. Although I did not overestimate what I had done, I felt very happy about the result, happy that my work in Princeton had begun in such a promising way. I went to Einstein to show him my proof. Einstein liked it very much, agreed that it was a great simplification and asked me to write it down for publication. He became more friendly, personal and enthusiastic.

"I should like it if you would work with me. We can have an exciting time together."

The same day I showed Einstein the proof he rang me up, asked me to come to see him. We walked through the garden of his house, and with perfect calmness, a calmness which I have never seen disturbed, he told me of the new ideas on which he was working. There were two problems which occupied his mind. One was that of constructing a representation of an elementary particle from the equations of the gravitational field, the most essential problem of the field theory. He had found a new method of approach and believed that the gravitational equations contain a regular solution, giving a strong field in a restricted part of a space which would then diminish rapidly. In other words, that it is possible to find a representation of "matter" in the equations of the gravitational field. In this belief Einstein was almost alone. He told me:

"I know that hardly any physicists believe that the gravitational forces can play any part in the constitution of matter. The physicist always argues that the forces are too small. This reminds me of a joke. An unmarried woman had a child and the family was greatly humiliated. So the midwife tried to console the mother by saying: 'Don't worry so much, it's a very small child!' " Playing with his hair, he began to laugh loudly at his joke.

The other problem was less spectacular but seemed to me very promising: the gravitational equations are very complicated, and we hardly know what treasures or disappointments they hide. Einstein had planned an approach to these equations from a new angle, to force them to betray some of their secrets. At that time he had his first ideas of what he later called "the new approximation method" for finding the solutions of the gravitational equations.

He said to me:

"I do not believe that I will live long. But I should like to spend the rest of my life on the problems which I regard as fundamental."

This attitude, presented with complete calm, appeared over and over again in our conversations. It seemed that the difference between life and death for Einstein consisted only in the difference between being able and not being able to do physics.

The most amazing thing about Einstein was his tremendous vital force directed toward one and only one channel: that of original thinking, of doing research. Slowly I came to realize that in exactly this lies his greatness. Nothing is as important as physics. No human relations, no personal life, are as essential as thought and the comprehension of how "God created the world." In this phrase so often repeated by Einstein with variations, was his peculiar religious feeling that laws of nature can be formulated simply and beautifully. When he had a new idea he asked himself: "Could God have created the world in this way?" or "Is this mathematical structure worthy of God?" Translated into ordinary language, this sentence means: "Is the theory logically simple enough?"

The next day I met Robertson in Fine Hall and told him:

"I am convinced now that gravitational waves do not exist. I believe I am able to show it in a very brief way."

Robertson was still skeptical:

"I don't believe you," and he suggested a more detailed discussion. He took the two pages on which I had written my proof and read it through.

"The idea is O.K. There must be some trivial mistake in your calculations."

He began quickly and efficiently to check all the steps of my argument, even the most simple ones, comparing the results on the blackboard with those in my notes. The beginning checked beautifully. I marveled at the quickness and sureness with which Robertson performed all the computations. Then, near the end, there was a small discrepancy. He got plus where I got minus.

We checked and rechecked the point; Robertson was right. At one place I had made a most trivial mistake, in spite of having repeated the calculations three times! Such a mistake must have as simple an explanation as a Freudian slip in writing or language. Subconsciously I did not believe in the result and my first doubts were still there. But I wanted to prove it in a simpler way than had Einstein. Thus I had to cheat myself. Such mistakes happen often, but they are usually caught before papers are printed. The author comes back many times to his work before he sees it in print and has, therefore, plenty of opportunity to gain a more detached attitude and to apply the criteria of logic which were repressed during the emotional strain of creation.

Thus my whole result went to pieces. Although it was a very small matter, I felt very downcast at the moment, as does every scientist in similar circumstances even though he has accustomed himself to disappointment. He feels the collapse of hope, the death of the creation of his brain. It is like the succession of birth and death.

Robertson tried to console me:

"It happens to everyone. The most trivial mistakes are always the most difficult to detect."

I was depressed over the incident. Certainly there was still the possibility that gravitational waves do not exist. My mistake only showed that my work was irrelevant to their existence or non-existence. But the discussions with Robertson convinced me that gravitational waves do exist. This conclusion seemed to follow from one of his previous results which he showed me. But if so, then there must be a mistake in Einstein's paper. The next day I went to Einstein and announced:

"My result was wrong. I made a mistake in calculating. I believe that the waves do exist."

Einstein said simply:

"I found a mistake in my paper last night. My proof is wrong too."

Einstein told me more about his mistake; it was less trivial than mine and difficult to detect. The dramatic aspect of the situation was emphasized by the fact that Einstein's lecture on his paper had been announced a week before in Fine Hall to be delivered just one day after he discovered his mistake. He had to lecture on his paper, but the paper did not prove anything.

In the afternoon of the same day the great lecture room in Fine Hall was crowded to the last seat. All lectures in the mathematical seminar are announced by a small card tacked to the bulletin board in the hall. The card contains the name of the lecturer and his subject. Einstein's lectures are the only exceptions. If there is a cryptic announcement: "On —— at 5 P.M. a mathematical seminar will be held in Room 113," practically everyone at Fine Hall knows that Einstein will be the lecturer. The announcement of his name would be sufficient to bring a crowd of journalists and photographers from New York.

I admired the skill, charm and honesty with which Einstein performed the difficult task of lecturing about a result which had been destroyed less than a day before. He lectured for forty-five minutes, stating the subject, the idea of the proof and the mistake which he had made. Finishing, he said:

"If you ask me whether there are gravitational waves or not, I must answer I do not know. But it is a highly interesting problem."

Later he came to the same conclusion to which I had come with Robertson's help: that waves do exist. His paper, which was later published in the *Bulletin of the Franklin Institute*, contains the correction of the mistake as well as a short statement concerning the changes through which the work went during the period of its preparation.

This story reveals the dramatic aspect of scientific work. After the ecstasy of creation painful doubts are climaxed by the confirmation or destruction of the result. Doubt is the worst! A negative result may be disagreeable, may force one to give up his belief in a theory, but at least it kills the problem and may indicate a new pathway. Even a decidedly negative result is a relief after a period of doubt.

The story also has its philosophical moral. During all our discussions we talked and worried about the gravitational waves: do they exist in nature or not? Einstein's and, I believe, every scientist's approach is that of a realist. There is nature, there are laws, and the scientist with his feeble mind tries to form a consistent, logically simple picture of reality, a picture which can explain the world of our sensual impressions. We did not know whether we had to create or destroy gravitational waves in this picture. For Einstein the gravitational waves did exist before he wrote his paper, then he destroyed them in his picture of reality and finally was forced to re-create them once more. The human mind creates, changes and re-creates its picture of the universe. But behind this activity, behind this wandering through mistakes, there is the basic belief in the existence of reality which the scientist tries to comprehend by his efforts. Do gravitational waves *really* exist or not? The sentence is meaningless. Nobody can formulate logically what this sentence means. But there is in the scientist the feeling that his efforts lead to a better and deeper understanding of nature. Every scientist is emotionally a realist. He does not think that *he* created or destroyed gravitational waves. He believes that he discovered their existence or nonexistence in nature. This belief, though impossible to formulate in a rational way, is one of the driving forces of all creation. Philosophical idealism is sterile. A mind which thinks that gravitational waves are or are not radiating only from his own brain cannot bother seriously with this problem. A scientist who has done research successfully and regards himself as an idealist must have acted in the moments of creation as a realist does, accepting emotionally the reality of the outside world and later building an artificial philosophical structure detached from the moments of his work and strange to its spirit, plunging himself into a dangerous discrepancy between deed and thought. In our everyday actions, in worrying about the health of our children, about the faithfulness of our wives, or while doing scientific research, we act always as realists. This realistic feeling toward the universe in which he is placed is so strong in Einstein that it takes the form of something appearing just the opposite. While talking about God and His creation of the world he means always the inner consistency and the logical simplicity of the laws of nature. I would call it "the realist's approach to God."

Einstein uses his concept of God more often than a Catholic priest. Once I asked him:

"Tomorrow is Sunday. Do you want me to come to you, so we can work?"

"Why not?"

"Because I thought perhaps you would like to rest on Sunday."

Einstein settled the question by saying with a loud laugh: "God does not rest on Sunday either."

In Fine Hall there is a room usually closed, but opened when a distinguished visitor is being entertained. Engraved on the fireplace is a sentence of Einstein's:

"*Rafiniert ist Herr Gott aber boshaft ist Er nicht,*" which someone in Fine Hall has translated into American:

"God is slick but He ain't mean."

V

IN AMERICA I saw for the first time in my life Negro dances and plays which were full of fire and vital force. The Savoy dance hall in Harlem changes into an African jungle with burning sun and richly growing vegetation. The air is full of vibration. Vital force emanates from the loud music and the passionate dancing until the whole atmosphere becomes unreal. In contrast the white people look half alive, ridiculous and humiliated. They help to form the background against which the primitive, unbounded vitality of the Negroes shines more brightly. One feels that any pause, any interval is unnecessary, that this intensive motion could go on forever.

I often had this picture in mind while watching Einstein work. There is a most vital mechanism which constantly turns his brain. It is the sublimated vital force. Sometimes it is even painful to watch. Einstein may speak about politics, listen kindly to requests and answer questions properly, but one feels behind this external activity the calm, watchful contemplation of scientific problems, that the mechanism of his brain works without interruption. It is a constant motion which nothing can stop. Other scientists have a switch which allows them to turn off or at least to decelerate the mechanism by a detective story, exciting parties, sex or a movie. There is no such switch in Einstein's brain. The mechanism is never turned off.

After the work on gravitational waves was ended Einstein began work immediately on other problems. I myself was discouraged by the bad start in these new and difficult surroundings and did not feel like returning to the problem of gravitation. For a short time I turned to the unitary field theory on which I had worked with Born, having found that it was possible to deepen it in one direction and to remove, at least partially, some of the objections which had been formulated by Einstein and others. I worked, though not with much fire, and obtained some results.

My collaboration with Einstein started a few weeks later. At the beginning there was no great enthusiasm on my part. Only later was I carried away by Einstein's power of reasoning, and I found myself in the midst of attractive, deep and interesting problems. For the next three years my scientific work was concentrated on one problem which can be formulated simply. It is the problem of motion. Its final solution will never influence our daily lives and will never have any technical application. It is a purely abstract problem. The goal is to gain a better understanding of the laws of motion, to formulate them more fully and more logically than in Newtonian mechanics. It is a basic problem rooted in the foundations of physics; I found it more exciting the longer I worked on it.

Again the underlying idea may sound simple, but to carry it out required the development of special techniques of calculation, more intensive thought and tedious work than any problem I had tackled before. For three years I worked on this problem whose only practical application that I know of is the analysis of the motion of double stars by methods giving deeper insight than the old Newtonian mechanics. For three years I have been bothering with double stars without ever having seen any!

The problem of motion is as old as human thought. In Newtonian mechanics there is always the concrete picture: the horse pulls a carriage and the carriage moves; a force accelerates the body on which it is acting. This picture of motion seems simple from the mechanical view only because we are accustomed to it. In this picture we have both the force and the particle, the basic concepts in the mechanistic point of view, well-known accessories of the old physics. But can we reconcile this picture with our new field concepts which have proved to be so successful elsewhere? We must analyze the problem of motion with the help of the concepts and methods of reasoning introduced by the field theory.

And this proved to be a difficult task. Neither in Maxwell's electromagnetic theory nor in Einstein's gravitational theory did the problem of motion at that time fit the field views. The problem of motion had not emerged from its mechanical frame. It was the purpose of our investigation to disregard the intuitive concepts of pushing, pulling and drawing, to formulate the problem in terms of the field theory and to solve it by using only the equations of the field. The problem is important from the point of view of principle and from the philosophical, but its practical value is nil.

On problems of such character, apparently completely detached from our everyday life, scientists have worked for generations. But on the average, in their final outcome, the efforts of the scientists have a utilitarian value satisfying existing needs or creating new ones. But I do not think that the conscious thought of utilitarian value has played an important role in a really great discovery. The urge to think, the emotions of creation, the attempts to escape from the dullness of everyday life form the motive force for scientific work.

No one has expressed this thought more beautifully than Einstein himself, when he said in one of his addresses:

. . . I believe with Schopenhauer that one of the strongest motives that lead men to art and science is to escape from everyday life, with its painful crudity and hopeless dreariness, from the fetters of one's own ever-shifting desires. A finely tempered nature longs to escape from personal life into the world of objective perception and thought; this desire may be compared with the townsman's irresistible longing to escape from his noisy, cramped surroundings into the silence of high mountains, where the eye ranges freely through the still, pure air and fondly traces out the restful contours which look as if they were built for eternity. With this negative motive there goes a positive one. Man tries to make for himself, in the way that suits him best, a simplified and intelligible picture of the world and thus to overcome the world of experience, for which he tries to some extent to substitute this cosmos of his. This is what the painter, the poet, the speculative philosopher and the natural scientist do, each in his own fashion. He makes this cosmos and its construction the pivot of his emotional life, in order to find in this way the peace and security which he cannot find in the narrow whirlpool of personal experience.

I do not believe that there are more than ten people in the world who have studied our papers on the problem of motion. How is it possible that publications on which Einstein's name appears, important though difficult papers, are read only by a handful of people? The explanation lies in Einstein's peculiar position in the modern scientific world. It is easy to say that Einstein is the greatest living physicist. But the sentence conveys little and satisfies only the newspapers when they discover that the number-one place among physicists belongs to Einstein and that one more *extremum* problem is solved. But to determine Einstein's position among scientists is much less simple.

The clue to the understanding of Einstein's role in science lies in his loneliness and aloofness. In this respect he differs from all other scientists I know. Perhaps Dirac is the nearest to Einstein, although the difference between them is still great. When, in 1905, Einstein formulated the special relativity theory his name was unknown to the world of science. He never had studied physics at a famous university, he was not attached to any school; he worked as a clerk in a patent office. Einstein once told me:

"Until I was almost thirty I never saw a real theoretical physicist."

I was tempted to ask him:

"Why didn't you look in the mirror?"

Nowadays, to learn the scientific technique, to be in contact with masters, to go through a good school of physics, to learn the use of the proper tools, is essential for every scientist. Thus the example of Einstein is unique. For him the isolation was a blessing since it prevented his thought from wandering into conventional channels. This aloofness, this independent thought on problems which Einstein formulated for himself, not marching with the crowd but looking for his own lonely pathways, is the most essential feature of his creation. It is not only originality, it is not only imagination, it is something more, which can be understood by a glimpse at the problems and methods of Einstein's work.

When Einstein was twenty-seven his scientific achievement was not confined to the theory of relativity. Fundamental ideas concerning Brownian motion and the theory of photons are connected with Einstein's name. All these results were recognized fairly soon, and his name became famous among physicists. This was in 1908. At that time physicists were not greatly interested in the problem of gravitation. The bulk of important scientific research lay in the quantum theory and special relativity theory, which arose from grave difficulties in the realm of electrodynamics. All results in quantum and special relativity theory were regarded as dramatic, modern and worthy of attention. But nobody saw any serious problem in gravitational theory. Calculations based on Newton's law of gravitation suited the astronomers. I do not think that in the maze of publications there was one important paper attempting to solve the basic problem of gravitation. Einstein devoted ten years of his life to this problem when no one else was interested in it. The special relativity theory removed a painful difficulty in physics and a vital contradiction. It was accepted by the world of physicists with enthusiasm. But toward the general relativity theory which tried to solve the problem of gravitation the attitude was "who cares?" Einstein told me of the lack of interest in this problem, how no one believed in the success of his method of attack, which seemed shockingly unconventional and strange. To ponder on a problem for ten years without any encouragement from the outside requires strength of character. This strength of character, perhaps more than his great intuition and imagination, led to Einstein's scientific achievements.

It was not until later that the sweeping success of the general relativity theory came. Fame among physicists became a worldwide fame greater than any other scientist had known.

Today Einstein still works on problems in which only a very few physicists are interested. In the meantime the general interest in physics has shifted several times, from the old quantum theory initiated by Planck and Einstein to the quantum theory of the atom following Bohr's fundamental paper which appeared in 1913. Then from the quantum theory of the atom to the wave or quantum mechanics initiated and developed by De Broglie, Schroedinger, Heisenberg, Dirac and Born. Then from the quantum mechanics to nuclear physics.

During this period of twenty-five years Einstein remained faithful to his methods of work, formulating his own problems and following his own path of thought. No one can foresee whether the old success will be repeated again, whether the solitary work will bring results as fundamental and essential as the general relativity theory. Very few of the younger generation of physicists are seriously interested in the problems with which Einstein occupies his life. Most of them work in close contact, gathering material, searching for theories, often of a provisional character, to fit the tremendous richness of experimental data in the realm of nuclear physics.

Einstein does not like the spirit of "engineering physics," looking for quick results, fitting a restricted domain of facts by an *ad hoc* invented hypothesis. His interest is, and will remain, in fundamental problems. There are not many physicists who share his taste. The majority is interested in obtaining results which satisfy the more momentary needs, in looking for theories and methods which often come to life and die as quickly as spring flowers. A book on nuclear physics, for example, cannot be modern nowadays because it takes a finite time interval to print it.

Einstein himself regards the solution of the problem of gravitation, presented in the general relativity theory, as the greatest scientific achievement of his life. He said to me:

"The special relativity theory would have been discovered by now whether I had done it or not. The problem was ripe. But I do not believe that this is true in the case of general relativity theory."

By this sentence I believe that Einstein meant to stress that the interests of physicists lay far away from the problems which the general relativity theory tackled and solved.

Einstein's habit of working everything independently is carried to the extreme. Once when we had to perform a calculation which was quoted in many books I suggested:

"Let us look it up. It will save time."

But he proceeded with his calculations, saying:

"It will be quicker this way. I have forgotten how to use books."

Before we published our paper I suggested to Einstein that I should look up the literature to quote scientists who had worked on this subject before. Laughing loudly, he said:

"Oh yes. Do it by all means. Already I have sinned too often in this respect."

VI

To the problem of motion on which we worked Einstein had turned again and again during the previous fifteen years. In the main the problem is this:

The general relativity theory formulates equations of the

gravitational field which describe changes of this field in space and time. From the point of view of the field theory, matter may be regarded as placed in regions where the equations of the field break down, where they are not valid any more. The essential task of the field equations is to describe the field, extending between particles, between the regions in which the gravitational equations break down. But what of the motion of the particles? It was assumed in the general relativity theory that this motion follows a special law, the so-called law of the "geodetic line." We do not need to concern ourselves with the specific meaning of these words. It is enough to understand that the mathematical structure of the theory of relativity contains equations of two kinds: (1) equations of the gravitational field, (2) equations of motion. This fusion of the two kinds of equations must be regarded as logically unsatisfactory. In the field equations we attempt to emerge from the old mechanical view; in the equations of motion we return to the errors of the old view. How can we remove this difficulty? Is it possible that the equations of motion are already contained in the field equations but that we are unable to deduce them? It may be that the field equations force the regions in which they break down—that is to say, the particles—to move along definite lines. It may be that the field equations impose restrictions upon the motion of the particles and that the additional law of motion is unnecessary and perhaps wrong. It may be that the old law of motion is only a first approximation of a deeper law which can be deduced from the field equations when one understands their contents. Einstein believed that the equations of motion are contained in the field equations and for years looked for a proof of this statement which would considerably simplify the logical structure of the general relativity theory.

The philosophical aspect of this work is not new. The simpler our assumptions are from the logical point of view, the longer is the chain of reasoning leading from the principal assumptions to results which can be checked or disproved by observation. Paradoxically enough, modern physics seems difficult and complicated because it is so simple. It seems difficult and complicated because we must start from fundamental, abstract-sounding arguments to travel a long way through complicated reasoning, link by link, in order to form the chain connecting our assumptions with observation.

This general trend can be seen clearly in our problem of motion. We began by assuming two types of laws: those of field and of motion. To deduce the law of motion in any *special* case was simple, because the *general* law of motion was assumed. But let us disregard the general law of motion, assuming that it is contained already in the field equations. Then we obtain a system which is logically much simpler. One system of equations instead of two! But the logical chain is greatly lengthened. We must now deduce the equations of motion from the field equations, and this is not an easy task. We pay with great technical complication for the logical simplification achieved. Nevertheless it is exactly the logical aspect which is essential for the scientist. He is ready to pay any price in computation to obtain a theory which is cleaner and purer from the logical point of view.

Since I should like to quote one of Einstein's beautiful phrasings, I must introduce some mathematical terms. The physical field theories, like Maxwell's theory of electrodynamics or Einstein's theory of gravitation, formulate equations called "partial differential equations." But to find a solution of these equations, to deduce results which can be compared with observation, we must apply a process called "integration." This is often a tedious and difficult procedure. The logical chain from theory to observation consists mostly of this procedure of "integrating" differential equations. Here lie our greatest technical difficulties.

When we had toiled for months over problems of this character Einstein used to remark:

"God does not care about our mathematical difficulties; He integrates empirically."

Here again is his belief that it is possible to reduce the laws of nature to simple principles and that their simplicity, and not the technical difficulties, forms a criterion of the beauty of our theories.

The problem of logically simplifying the theory of gravitation by disregarding the equations of motion proved to be difficult. After crossing out the equations of motion the remaining field equations were very complicated, and we knew little of what they might hide. Mathematical physicists had scratched their surface but were ignorant of their deeper content. We did not know whether they imposed restrictions on the motion of particles or whether they allowed them arbitrary motion. In other words, we did not know whether equations of motion must be or need not be added to the field equations.

Before we began our thorough investigation of this problem Einstein already had his idea of a new approach to the field equations by applying a "new approximation method." He hoped to achieve two results: first, to show that the equations of motion are contained in the field equations; second, to find a hidden treasure which would allow us to build a bridge from the classical gravitational theory to quantum theory. Einstein believed that from the laws governing the stars and planets we could deduce laws governing the inside of the atom.

He thought out a very ingenious approach to this wide field of research and suggested that I work with him. His aim was one that he had kept clearly before him for a long time: to obtain the equations of motion and to find a connection between classical and quantum physics.

I was not convinced at the beginning that the equations of motion are contained in the field equations. But I was still more skeptical about the second aim, refusing to believe that there is any connection between the gravitational and quantum theories. It seems presumptuous that I would dare to differ with Einstein on any subject, but I know that there is nothing so dangerous in science as blind acceptance of authorities and dogmas. My own mind must remain for me the highest authority. Nearly every understanding is gained by a painful struggle in which belief and disbelief are dramatically interwoven. I wanted to make this point quite clear to Einstein:

"If I once take for granted that you must always be right, then there is nothing for me to do but to nod my head and perform the mechanical calculations. The whole fun of scientific work will be gone. I know that in the beginning my skepticism may be annoying and you may find little help in my collaboration."

Looking back on it now, I must admire the patience with which Einstein treated my objections. When we started he was far ahead of me in this problem and I had difficulty in following him. But he was never impatient; he came back many times to the same explanation of ways and methods, considered all my doubts seriously until I had absorbed the principal idea. Once he made a remark which I considered to be a great compliment:

"I know your character very well because I am just the same. I do not believe anyone until I understand everything for myself."

Einstein had a difficult time with me at the start. Instead of proving that the equations of motion are contained in the field equations I tried for a long time to prove just the opposite. Einstein did little to influence me. He said later:

"If I had bullied you, it would have taken still longer to convince you that you were looking in the wrong direction."

But one day I saw the light. All my skepticism vanished and I began to collaborate with great enthusiasm. Once I believed in the result my attitude changed radically. But there was still skepticism and disbelief in one direction. I was convinced by then that the equations of motion are contained in the field equations, but I did not believe that the problem had anything to do with quantum theory. Einstein convinced me on the first point, but with respect to the second I stressed the difference of our opinions as freely and directly as is possible only with such a man as Einstein.

Thus the time of our close collaboration began, a time filled with tension and excitement. Our work split into two distinct parts: the first largely intricate reasoning and presenting the general theory, the second marked by special, tedious calculations leading to the explicit formulation of the equations of motion torn out from the field equations.

We saw each other practically every afternoon, nearly always at Einstein's, sometimes at my home.

We worked in Einstein's study on the second floor of his house; the first floor had become a hospital during his wife's illness. At that time, I believe, all hope for her recovery had been lost. I know for a fact that Einstein gave his wife the greatest care and sympathy. But in this atmosphere of coming death Einstein remained serene and worked constantly. A few days after his wife's death I learned that Einstein was again spending the mornings in Fine Hall. I went to see him. He looked tired, his complexion was more sallow than before. I pressed his hand but could not bring myself to say the trivial words of sympathy. We discussed a serious difficulty in our scientific problem as though nothing had happened. Einstein worked with equal intensity during his wife's illness as later, after she died. There is no force which can stop Einstein's work as long as he is alive.

Although Einstein is most understanding, patient and kind, the collaboration was not easy. The reason lay in the richness of his ideas, in the fact that he was always ahead of me, forcing me, by his own constant accomplishment, to a continuous state of excited activity. Not to be left behind I had to keep working, trying to solve the serious difficulties which we encountered while collaborating on the problem of motion.

Sometimes after we separated I would think in the night about our last discussion, and a new idea would strike me, illuminating the subject from a new angle. Next day I would rush to Einstein, often only to find that he had come to the same conclusion and was still further along. During our collaboration on the general part of the problem, which proved difficult and required many ideas, my contribution concerned one essential aspect only. I furnished the proof that the problem of motion can throw no light on the quantum theory. Here my skepticism won. It guided me to the proof, which was amazingly simple. It was interesting to see Einstein trying to find a fault in my reasoning. After checking the links of thought he admitted that it was right but next day grew suspicious again and began to analyze each step most carefully. But the proof held. He said:

"Yes, I am now convinced that we cannot obtain quantum restrictions for motion from the gravitational equations."

I asked him:

"Won't you change your mind tomorrow?"

Stretching his hand to me, he said with exaggerated earnestness:

"No, never. I give you my hand on it," and he burst into his loud laugh.

With this one exception the other essential ideas were Einstein's.

There remained the problem of special calculations, which were very difficult and required a technique which had to be especially developed for this purpose. Here Einstein trusted me completely and took little part in the work. He never made any attempt to check the calculations but was very much interested in discussing all the difficulties which arose. The computations were checked and rechecked many times, not without finding deep as well as superficial errors, until finally the equations of motion for a two-body problem were established.

During this time I was in close contact with Robertson and told him how our work was progressing. It was a pleasure to discuss physics with him. I learned to admire his quick, well-ordered mind in which everything was fitted neatly into the right compartment, his keen intuition on scientific matters, his great vitality and robust sense of humor. Our friendship grew, and at the time I obtained the equations of motion we already knew each other well.

It was Robertson who deduced the final consequences of our theory by solving the differential equations of motion and by finding explicitly the laws governing the motion of double stars according to the general relativity theory.

From Newton's theory it follows that the earth, or any other planet, moves along an ellipse with the sun at one of its foci. The result is slightly more complicated for a double star, where we cannot assume that the mass of one of them is much greater than that of the other, as we do in the case of the sun-planet problem. According to Newtonian mechanics, each of the two stars moves along an ellipse and the common focus of the ellipses is the so-called "center of gravity" of the double star.

This result, regarded as rigorous by the Newtonian theory, forms only a first approximation from the point of view of the theory of relativity. There is always some truth in an old theory, and its limitations are shown by the broader view gained through the climb upward to the new theory. The old theory is replaced by the new one either because the new theory embraces a wider range of facts or because the new theory is logically simpler or because of both these reasons. It keeps its place in physics until again our view broadens or until the need for a wider and logically simpler theory arises in the actual progress of science.

The connection between the old and new theory appeared clearly in our case of motion. In its more crude features the final result was the same as in Newtonian mechanics. In a different, roundabout way we obtained, in the first approximation, the same results as in the old physics: elliptic motion of the double stars. The new way was much more difficult and complicated because the basic principles were simpler. But there was one more difference between the old and new approach, characteristic of the transition from old to new theory. The Newtonian equations of motion were obtained from our theory as the first approximation. We asked: what equations, different from the Newtonian, shall we obtain by pushing our approximation method one step further? After months of calculation we established for double stars more exact, more accurate equations of motion than those formulated in Newtonian mechanics. These were the equations which I gave to Robertson and from which he made the final deductions. The only essential difference in the conclusions drawn from the old and new equations lay in the so-called perihelion motion. Not only does each of the stars move along an ellipse, but the whole ellipse turns through a complete rotation in millions of years.

VII

THE PROGRESS OF MY WORK with Einstein brought an increasing intimacy between us. More and more often we talked of social problems, politics, human relations, science, philosophy, life and death, fame and happiness and, above all, about the future of science and its ultimate aims. Slowly I came to know Einstein better and better. I could foresee his reactions; I understood his attitude which, although strange and unusual, was always fully consistent with the essential features of his personality.

Seldom has anyone met as many people in his life as Einstein has. Kings and presidents have entertained him; everyone is eager to meet him and to secure his friendship. It is comparatively easy to meet Einstein but difficult to know him. His mail brings him letters from all over the world which he tries to answer as long as there is any sense in answering. But through all the stream of events, the impact of people and social life forced upon him, Einstein remains lonely, loving solitude, isolation and conditions which secure undisturbed work.

A few years ago, in London, Einstein made a speech in Albert Hall on behalf of the refugee scientists, the first of whom had begun to pour out from Germany all over the world. Einstein said then that there are many positions, besides those in universities, which would be suitable for scientists. As an example he mentioned a lighthouse keeper. This would be comparatively easy work which would allow one to contemplate and to do scientific research. His remark seemed funny to every scientist. But it is quite understandable from Einstein's point of view. One of the consequences of loneliness is to judge everything by one's own standards, to be unable to change one's co-ordinate system by putting oneself into someone else's being. I always noticed this difficulty in Einstein's reactions. For him loneliness, life in a lighthouse, would be most stimulating, would free him from so many of the duties which he hates. In fact it would be for him the ideal life. But nearly every scientist thinks just the opposite. It was the curse of my life that for a long time I was not in a scientific atmosphere, that I had no one with whom to talk physics. It is commonly known that stimulating environment strongly influences the scientist, that he may do good work in a scientific atmosphere and that he may become sterile, his ideas dry up and all his research activity die if his environment is scientifically dead. I knew that put back in a gymnasium, in a provincial Polish town, I should not publish anything, and the same would have happened to many another scientist better than I. But genius is an exception. Einstein could work anywhere, and it is difficult to convince him that he is an exception.

He regards himself as extremely lucky in life because he never had to fight for his daily bread. He enjoyed the years spent in the patent office in Switzerland. He found the atmosphere more friendly, more human, less marred by intrigue than at the universities, and he had plenty of time for scientific work.

In connection with the refugee problem he told me that he would not have minded working with his hands for his daily bread, doing something useful like making shoes and treating physics only as a hobby; that this might be more attractive than earning money from physics by teaching at the university. Again something deeper is hidden behind this attitude. It is the "religious" feeling, bound up with scientific work, recalling that of the early Christian ascetics. Physics is great and important. It is not quite right to earn money by physics. Better to do something different for a living, such as tending a lighthouse or making shoes, and keep physics aloof and clean. Naïve as it may seem, this attitude is consistent with Einstein's character.

I learned much from Einstein in the realm of physics. But what I value most is what I was taught by my contact with him in the human rather than the scientific domain. Einstein is the kindest, most understanding and helpful man in the world. But again this somewhat commonplace statement must not be taken literally.

The feeling of pity is one of the sources of human kindness. Pity for the fate of our fellow men, for the misery around us, for the suffering of human beings, stirs our emotions by the resonance of sympathy. Our own attachments to life and people, the ties which bind us to the outside world, awaken our emotional response to the struggle and suffering outside ourselves. But there is also another entirely different source of human kindness. It is the detached feeling of duty based on aloof, clear reasoning. Good, clear thinking leads to kindness and loyalty because this is what makes life simpler, fuller, richer, diminishes friction and unhappiness in our environment and therefore also in our lives. A sound social attitude, helpfulness, friendliness, kindness, may come from both these different sources; to express it anatomically, from heart and brain. As the years passed I learned to value more and more the second kind of decency that arises from clear thinking. Too often I have seen how emotions unsupported by clear thought are useless if not destructive.

Here again, as I see it, Einstein represents a limiting case. I had never encountered so much kindness that was so completely detached. Though only scientific ideas and physics really matter to Einstein, he has never refused to help when he felt that his help was needed and could be effective. He wrote thousands of letters of recommendation, gave advice to hundreds. For hours he talked with a crank because the family had written that Einstein was the only one who could cure him. Einstein is kind, smiling, understanding, talkative with people whom he meets, waiting patiently for the moment when he will be left alone to return to his work.

Einstein wrote about himself:

My passionate interest in social justice and social responsibility has always stood in curious contrast to a marked lack of desire for direct association with men and women. I am a horse for single harness, not cut out for tandem or teamwork. I have never belonged wholeheartedly to country or state, to my circle of friends or even to my own family. These ties have always been accompanied by a

vague aloofness, and the wish to withdraw into myself increases
with the years.

Such isolation is sometimes bitter, but I do not regret being cut
off from the understanding and sympathy of other men. I lose
something by it, to be sure, but I am compensated for it in being
rendered independent of the customs, opinions and prejudices of
others and am not tempted to rest my peace of mind upon such
shifting foundations.

For scarcely anyone is fame so undesired and meaningless as
for Einstein. It is not that he has learned the bitter taste of fame,
as frequently happens, after having desired it. Einstein told me
that in his youth he had always wished to be isolated from the
struggle of life. He was certainly the last man to have sought
fame. But fame came to him, perhaps the greatest a scientist has
ever known. I often wondered why it came to Einstein. His ideas
have not influenced our practical life. No electric light, no telephone, no wireless is connected with his name. Perhaps the only
important technical discovery which takes its origin in Einstein's theoretical work is that of the photoelectric cell. But
Einstein is certainly not famous because of this discovery. It is
his work on relativity theory which has made his name known
to all the civilized world. Does the reason lie in the great influence of Einstein's theory upon philosophical thought? This
again cannot be the whole explanation. The latest developments
in quantum mechanics, its connection with determinism and indeterminism, influenced philosophical thought fully as much.
But the names of Bohr and Heisenberg have not the glory that
is Einstein's. The reasons for the great fame which diffused
deeply among the masses of people, most of them removed from
creative scientific work, incapable of estimating his work, must
be manifold and, I believe, sociological in character. The explanation was suggested to me by discussions with one of my
friends in England.

It was in 1919 that Einstein's fame began. At this time his great
achievement, the structure of the special and general relativity
theories, was essentially finished. As a matter of fact it had been
completed five years before. One of the consequences of the

general relativity theory may be described as follows: if we photograph a section of the heavens during a solar eclipse and the same section under normal conditions, we obtain slightly different pictures. The gravitational field of the sun slightly disturbs and deforms the path of light, therefore the photographic picture of a section of the heavens will vary somewhat during the solar eclipse from that under normal conditions. Not only qualitatively but quantitatively the theory of relativity predicted the difference in these two pictures. English scientific expeditions sent in 1919 to different parts of the world, to Africa and South America, confirmed this prediction made by Einstein.

Thus began Einstein's great fame. Unlike that of film stars, politicians and boxers, the fame persists. There are no signs of its diminishing; there is no hope of relief for Einstein. The fact that the theory predicted an event which is as far from our everyday life as the stars to which it refers, an event which follows from a theory through a long chain of abstract arguments, seems hardly sufficient to raise the enthusiasm of the masses. But it did. And the reason must be looked for in the postwar psychology.

It was just after the end of the war. People were weary of hatred, of killing and international intrigues. The trenches, bombs and murder had left a bitter taste. Books about war did not sell. Everyone looked for a new era of peace and wanted to forget the war. Here was something which captured the imagination: human eyes looking from an earth covered with graves and blood to the heavens covered with stars. Abstract thought carrying the human mind far away from the sad and disappointing reality. The mystery of the sun's eclipse and the penetrating power of the human mind. Romantic scenery, a strange glimpse of the eclipsed sun, an imaginary picture of bending light rays, all removed from the oppressive reality of life. One further reason, perhaps even more important: a new event was predicted by a *German* scientist Einstein and confirmed by *English* astronomers. Scientists belonging to two warring nations had collaborated again! It seemed the beginning of a new era.

It is difficult to resist fame and not to be influenced by it. But fame has had no effect on Einstein. And again the reason lies in his internal isolation, in his aloofness. Fame bothers him when and as long as it impinges on his life, but he ceases to be conscious of it the moment he is left alone. Einstein is unaware of his fame and forgets it when he is allowed to forget it.

Even in Princeton everyone looks with hungry, astonished eyes at Einstein. During our walks we avoided the more crowded streets to walk through fields and along forgotten byways. Once a car stopped us and a middle-aged woman got out with a camera and said, blushing and excited:

"Professor Einstein, will you allow me to take a picture of you?"

"Yes, sure."

He stood quiet for a second, then continued his argument. The scene did not exist for him, and I am sure after a few minutes he forgot that it had ever happened.

Once we went to a movie in Princeton to see the *Life of Émile Zola.* After we had bought our tickets we went to a crowded waiting room and found that we should have to wait fifteen minutes longer. Einstein suggested that we go for a walk. When we went out I said to the doorman:

"We shall return in a few minutes."

But Einstein became seriously concerned and added in all innocence:

"We haven't our tickets any more. Will you recognize us?"

The doorman thought we were joking and said, laughing:

"Yes, Professor Einstein, I will."

Einstein is, if he is allowed to be, completely unaware of his fame, and he furnishes a unique example of a character untouched by the impact of the greatest fame and publicity. But there are moments when the aggressiveness of the outside world disturbs his peace. He once told me:

"I envy the simplest working man. He has his privacy."

Another time he remarked:

"I appear to myself as a swindler because of the great publicity about me without any real reason."

Einstein understands everyone beautifully when logic and thinking are needed. It is much less easy, however, where emotions are concerned; it is difficult for him to imagine motives and emotions other than those which are a part of his life. Once he told me:

"I speak to everyone in the same way, whether he is the garbage man or the president of the university."

I remarked that this is difficult for other people. That, for example, when they meet him they feel shy and embarrassed, that it takes time for this feeling to disappear and that it was so in my case. He said:

"I cannot understand this. Why should anyone be shy with me?"

If my explanation concerning the beginning of Einstein's fame is correct, then there still remains another question to be answered: why does this fame cling so persistently to Einstein in a changing world which scorns today its idols of yesterday? I do not think the answer is difficult.

Everything that Einstein did, everything for which he stood, was always consistent with the primary picture of him in the minds of the people. His voice was always raised in defense of the oppressed; his signature always appeared in defense of liberal causes. He was like a saint with two halos around his head. One was formed of ideas of justice and progress, the other of abstract ideas about physical theories which, the more abstruse they were, the more impressive they seemed to the ordinary man. His name became a symbol of progress, humanity and creative thought, hated and despised by those who spread hate and who attack the ideas for which Einstein's name stands.

From the same source, from the desire to defend the oppressed, arose his interest in the Jewish problem. Einstein himself was not reared in the Jewish tradition. It is again his detached attitude of sympathy, the rational idea that help must be given where help is needed, that brought him near to the Jewish problem. Jews have made splendid use of Einstein's gentle attitude. He once said:

"I am something of a Jewish saint. When I die the Jews will take my bones to a banquet and collect money."

In spite of Einstein's detachment I had often the impression that the Jewish problem is nearer his heart than any other social problem. The reason may be that I met him just at the time when the Jewish tragedy was greatest and perhaps, also, because he believes that there he can be most helpful.

Einstein also fully realized the importance of the war in Spain and foresaw that on its outcome not only Spain's fate but the future of the world depended. I remember the gleam that came into his eyes when I told him that the afternoon papers carried news of a Loyalist victory.

"That sounds like an angel's song," he said with an excitement which I had hardly ever noticed before. But two minutes later we were writing down formulae and the external world had again ceased to exist.

It took me a long time to realize that in his aloofness and isolation lie the simple keys leading to an understanding of many of his actions. I am quite sure that the day Einstein received the Nobel prize he was not in the slightest degree excited and that if he did not sleep well that it was because of a problem which was bothering him and not because of the scientific distinction. His Nobel prize medal, together with many others, is laid aside among papers, honorary degrees and diplomas in the room where his secretary works, and I am sure that Einstein has no clear idea of what the medal looks like.

Einstein tries consciously to keep his aloofness intact by small idiosyncrasies which may seem strange but which increase his freedom and further loosen his ties with the external world. He never reads articles about himself. He said that this helps him to be free. Once I tried to break his habit. In a French newspaper there was an article about Einstein which was reproduced in many European papers, even in Poland and Lithuania. I have never seen an article which was further from the truth than this one. For example, the author said that Einstein wears glasses, lives in Princeton in one room on the fifth floor, comes to the institute at 7 A.M., always wears black, keeps many of his

technical discoveries secret, etc. The article could be character-
ized as the apex of stupidity if stupidity could be said to have a
apex. Fine Hall rejoiced in the article and hung it up as a curiosity on the bulletin board at the entrance. I thought it so funny
that I read it to Einstein, who at my request listened carefully
but was little interested and refused to be amused. I could see
from his expression that he failed to understand why I found it
so funny.

One of my colleagues in Princeton asked me:

"If Einstein dislikes his fame and would like to increase his
privacy, why does he not do what ordinary people do? Why
does he wear long hair, a funny leather jacket, no socks, no
suspenders, no collars, no ties?"

The answer is simple and can easily be deduced from his
aloofness and desire to loosen his ties with the outside world.
The idea is to restrict his needs and, by this restriction, increase
his freedom. We are slaves of millions of things, and our slavery
progresses steadily. For a week I tried an electric razor—and one
more slavery entered my life. I dreaded spending the summer
where there was no electric current. We are slaves of bathrooms,
Frigidaires, cars, radios and millions of other things. Einstein
tried to reduce them to the absolute minimum. Long hair mini-
mizes the need for the barber. Socks can be done without. One
leather jacket solves the coat problem for many years. Suspenders are superfluous, as are nightshirts and pajamas. It is a minimum problem which Einstein has solved, and shoes, trousers,
shirt, jacket, are the very necessary things; it would be difficult
to reduce them further.

I like to imagine Einstein's behavior in an unusual situation.
For example: Princeton is bombed from the air; explosives fall
over the city, people flee to shelter, panic spreads over the town
and everyone loses his head, increasing the chaos and fear by his
behavior. If this situation should find Einstein walking through
the street, he would be the only man to remain as quiet as before.
He would think out what to do in this situation; he would do it
without accelerating the normal speed of his motions and he
would still keep in mind the problem on which he was thinking.

There is no fear of death in Einstein. He said to me once: "Life is an exciting show. I enjoy it. It is wonderful. But if I knew that I should have to die in three hours it would impress me very little. I should think how best to use the last three hours, then quietly order my papers and lie peacefully down."

VIII

Fine hall is, I believe, the most luxurious building ever devoted to mathematics. Everything is brand new, clean and comfortable. Downstairs are two splendid lecture rooms. Around the first and second floors are offices for university and institute professors. Each of them is lavishly furnished: a sofa, a blackboard which can be closed like an altar into a piece of furniture, a big desk, comfortable armchairs, a telephone. With jealousy I looked at the splendid rooms. Since even in Fine Hall space is limited, Fellows do not have rooms. They must be satisfied with drawers in the library, the only place in Fine Hall where they can work.

In the commons each day tea is served between four-thirty and five. It is the focal point of all social activity. The air is full of mathematical ideas and formulae. You have only to stretch out your hand, close it quickly and you feel that you have caught mathematical air and that a few formulae are stuck to your palm. The stained-glass windows in the room contain formulae: one of them Newton's gravitational law, the second Einstein's principal equations of general relativity theory, the third Heisenberg's uncertainty relations concerning quantum mechanics. Even the sun rays must remember, when passing through the windows, the laws to which they are subject according to the will of God, Newton, Einstein and Heisenberg.

There are about fifty members of the mathematical school for advanced study and perhaps as many professors and graduate students from the university. All together there are about one hundred mathematicians doing research. It is possible to be lonely in this crowd and to have but a few acquaintances. But if one is looking for a job, as half of them are, the right acquaintances are extremely important. As a rule, contacts are made at teatime. It is an art to make contact with the important members and to be able to interest them, to keep their attention and avoid boring them. Obvious gambits, such as pedagogical problems and differences between European and American life, simply won't do. Fine Hall has its level and is proud of it.

To make the right acquaintances, to fascinate the right people, is essential to one's career. Scientific results are important but not important enough. Princeton does not forgive a disregard for the rules of the game.

People watch carefully to see who talks with whom during teatime. I knew one of the younger professors, a great snob, who changed his indifferent and stiff attitude toward me after he saw me talking for fifteen minutes with one of the most important professors.

I soon found out that I was unable to play the game. I wanted very much to remain in America. I wanted to find a job, as did many others, and I tried my best not to show that I was one of the queue. I met, we shall say, Professor X. I tried to be very charming and polite to him because I knew how important it was to make a good impression on him. When our conversation happened to touch scientific subjects I tried to show genuine interest, knowledge and understanding and to demonstrate them by my remarks and questions. Merely nodding was not enough. But how was it possible to be bright and clever when I was distracted by the constant stream of thoughts at the back of my mind?

"Don't be too polite and obvious. Professor X. meets hundreds like you and knows they are looking for jobs. Do it in a more subtle way. Are you not ashamed of yourself, you dirty little snob? Here you are, ready to agree on all subjects with X., to laugh loudly at the jokes which you have heard many times before, all because you are angling for a job. You think you are smart, but really you are artificial and stupid. You remember yesterday when you watched another man, looking for a job, talking with the same Professor X? It disgusted you. Did you like it? Certainly not. You saw through all his little tricks and fawning smiles. This is how you look now to the outside observer."

After a few experiences I could no longer bear the strain and seldom went to tea. I know that my reactions were probably not typical, that they were intensified by my peculiar position, by the fact that I was not a refugee, not an American, but still looking for a job, and that I was, despite my age, young in scientific experience.

The atmosphere there, apart from its scientific aspect, was quite different from that in European universities. Many of the younger people in Fine Hall were refugees from Germany or research workers from American universities on fellowships in Princeton. Nearly all of them were looking for permanent jobs, trying at the same time to make use of the splendid atmosphere in Fine Hall by doing research and improving their chances by gaining the support of influential people. This was bound to create tension. The word most used from February to May was the word "job." "He got a job." "You have a chance of a job." "I wrote an application for a job." "The job situation is bad this year."

Except for Jews and radicals, the academic career in Europe progressed in a slow, continuous way, nearly automatically. The Ph.D., the docentship, the professorship. And that was all. A professor was secure for the rest of his life. It is different here. There are great unexpected jumps in position and in space. From assistant professor at one of the Eastern universities to head of the department at a Western university, or again—and this was without parallel in Europe—instructors or even professors might lose their jobs without warning. In this atmosphere of change Fine Hall plays the role of a reservoir, furnishing other universities with mathematicians. It is greatly to the credit of the institute that it takes this part of its task seriously. Many influential members of the institute and university devote much time and energy to providing positions for young scientists. But this mixture of search for research and search for jobs creates tension. The teas at Fine Hall are scientific slave markets on a small scale. From time to time a head of a department or a dean from a distant university comes to Fine Hall to meet the right man. He states his terms, his demands, and is introduced to the man approximately filling them. Then interviews are arranged; the candidate collects all his papers and reprints, writes an application, spends weeks looking impatiently in his mail for an answer and is expected to do research in the meantime.

The meetings of the mathematical or physical society are slave markets on a much broader scale. One of the ways of getting a job is to attend the meetings, lecture on your papers and meet the kind eye of someone who wants to buy your services and pay the price. The difficulty is that the market is overcrowded and there are many who come to be sold but few who come to buy. At one meeting I met a friend with a permanent professorship. I said to him:

"You seem to be one of the very few who came here for the pure love of science."

To which he answered:

"I wouldn't be too sure about it."

Even well-established professors are looking for offers from other universities. It is a traditional method of mild blackmail in many universities, the only method by which advancement may be achieved. I once suggested to a professor:

"Why not organize a proper racket out of this custom?"

He said mysteriously:

"You're not the first to have this idea."

There are a great number of graduate lectures in Fine Hall. They are given by the professors of the university, professors of the institute, invited professors, some of the Fellows and occasional visitors. If one wants to see a famous mathematician one does not need to go to him; it is enough to sit quietly in Princeton, and sooner or later he must come to Fine Hall.

Not knowing which to choose among the many lectures, I attended none, with the exception of theoretical physics seminars. Since I started to work with Einstein at home weeks sometimes passed without my entering the famous building. The place began to depress me. The density of the mathematical air made breathing difficult. Instead of accustoming myself to it by forcing myself to go, I stayed away, which made accidental contacts still more difficult.

Sometimes I needed to do some research in the library. It is on the third floor in Fine Hall and by far the most beautiful and best-organized mathematical library I have ever seen. Provided lavishly with books, periodicals and good furniture, it is open day and night and has, I believe, the most wonderful and helpful librarian in the world.

I would start to go to the library, but before reaching the third floor I would meet someone on the stairs.

"Why are you here so seldom?"

"Because I am working."

"How is it going?"

"Fairly well."

Then it would be my turn to ask:

"And your work?"

"I have some difficulty. It is a very interesting problem. Would you like me to tell you about it?" And that would be that!

There are special rooms in Fine Hall with blackboards for discussion. We would enter one of these rooms and discuss the problem. Sometimes our conversation would turn out to be so interesting that it would be interrupted only by hunger. Sometimes the discussion would be about science, sometimes it would turn to jobs and gossip; usually it was a mixture of these topics. After finally escaping I would return to the library and peacefully begin to read the paper I needed. I would feel someone looking over my shoulder. Then I would hear a whisper:

"What do you think about that paper?"

"I just started it. Do you know it?"

"I looked through it."

"How did you like it?"

The librarian would look gravely at us. One of us would suggest going to a room for a few minutes' talk. Once we did this more hours would pass.

In this splendid scientific atmosphere of Fine Hall the reading of a simple paper takes incomparably more time than reading it at home. But through it all at least one knows that he is in a "splendid scientific atmosphere." The possibilities for the "free exchange of ideas" make it difficult to manufacture one's own ideas. Everything is perfect, perhaps too perfect. Nevertheless, if I ever have a sabbatical leave, a free year in which I can go where I like, I am sure I shall go to Princeton.

It is so much easier to make acquaintances and friends in America than in England. Someone told me:

"You come to England and you have no friends for a long time. Then you leave England after two years and you have a few good friends. You come to America and you immediately have many friends. You leave America after two years and you haven't any friends."

That part of the remark which refers to the ease in making friends here, when compared with England, is true. The rest is certainly exaggerated, but there is some truth in it. The social life in Princeton is broad and superficial. I made some friends, but mostly among younger people, with whom I was bound by our common struggling for existence. The only professors with whom I came in close contact were Einstein and Robertson.

It is difficult to learn anything about America in Princeton—much more so than to learn about England in Cambridge. In Fine Hall English is spoken with so many different accents that the resultant mixture is termed "Fine Hall English." But through talks, readings, meetings with scientists from all corners of this vast country, I began to sense its atmosphere. My European feeling of superiority quickly vanished. Here I met Robertson, one of the most intelligent and well-educated men I have ever known, who could in one breath quote Plato, discuss Chinese art and recite ribald limericks. Here I met Melbar, a woman theoretical physicist with broad interests outside her field of research, the first Gentile with whom I could discuss the Jewish problem freely. I began to sense the dynamic atmosphere of this great country, full of hope for the future, a country to which the center of civilization is steadily shifting and the country which, I hope, will preserve freedom of thought for the years to come.

In Poland, when talking to a non-Jew, I was always well aware of the difference between us. My thoughts formed a sequence:

"I am a Jew and he is a Gentile."

Then:

"He is well aware that he is a Gentile and that I am a Jew, and he feels his superiority."

Then one step further:

"I despise this superior-feeling Gentile who cannot forget for a moment that he is talking to a Jew."

And so it went further until that ever-present thought killed any possibility of free expression.

In America I felt the difference between the pressure here and in my native country. Pressure exists here too. But the difference is great, and through the convex mirror of my past experience I see the picture of the extent to which it can grow if not opposed. It was a relief for me to sense the increased freedom here and to find understanding and warm, human sympathy in non-Jews.

In this respect America gave me more than England, where warm human relations are difficult to achieve; it takes a long time to conquer the shyness created by English detachment and to bring the English people to the melting point of understanding.

It is much easier here. I remember my talks with Melbar. She told me:

"Until I came East I was unaware of the existence of the Jewish problem. In the beginning I was horrified when I heard people ask whether or not someone was a Jew. Then I was horrified again that Jews do the same with Gentiles. It never enters my head to formulate the question whether somebody is a Jew or not. Why can't you make up your mind to ignore this question, Ludwik?"

I believed her. Melbar, never raising her voice, always quiet, became upset and angry whenever the question of somebody being a Jew or Gentile arose. She achieved the highest possible freedom with respect to the Jewish problem; it never existed for her as a personal problem.

I thought Melbar was the sole exception. But I found the same attitude later in Helen, who helped me to see my accumulated bitterness, my hypersensitiveness, and to lessen the inner pressure and increase my freedom. I cannot claim to be completely free of the influences which have accumulated during the years. But I find that an atmosphere free of hate decreases tension, brings peace and a growing kindness toward my fellow men.

IX

I came to Princeton in October 1936. Five months later, in February, my work with Einstein was developing and our collaboration was well cemented by our complete understanding and partial success, but the problem of formulating the equations of motion was still far from a final solution.

February is the time of appointments for the next academic year. The temperature in Fine Hall rose and so did the frequency of the word "job" by which this rise could be measured. My colleagues talked nervously about the future, confiding in each other and asking for news. Walking to and from Fine Hall and looking for the mail was filled with the excitement of expectation or the depression of disappointment. One of my colleagues received an offer from a university in China, and after he had booked passage the university ceased to exist, destroyed by bombs. Another of my friends was given a fellowship in Sweden and three of them jobs in this country. I had no offer, although there was some talk of a position in China. My fellowship was coming to an end, my bank account was depleted. Loria wrote me not to return to Poland. I received a letter from the Polish Association of University Assistants stating that as a Jew I could no longer consider myself a member. Anti-Semitism in Poland was growing rapidly. Each letter which came from my country began with the sentence, "I envy you, being in America."

I hoped to obtain a fellowship from the institute for the next year. This was not unusual. Quite frequently fellowships were extended a second year, although two years was usually the maximum. I decided to talk to Einstein about my personal problems. It was not an easy decision, because I knew how many people asked for his help. I knew that Einstein would do whatever he could, but I also understood that the chances of his being able to help me were very small. Strange as it seems, Einstein's support often means much less than that of many incomparably less famous professors. In Princeton his influence was amazingly small. He told me:

"My fame begins outside Princeton. My word counts for little in Fine Hall."

But even outside Princeton Einstein's recommendation did not mean as much as one would expect. The reason for this again lay in his great kindness. He had signed so many letters of recommendation that his generosity became known, and the letters were kept as valuable autographs rather than as letters of recommendation. I heard a true story concerning an opening in a hospital for an X-ray physicist, for which several refugees applied. Each had a recommendation from Einstein. I told Einstein this story and asked him whether it were true. It demonstrated for me that exaggerated kindness defeats its own purpose, and I wanted to make this point clear. But Einstein disagreed:

"I recommended four physicists but each one for a different reason, and I stated the reasons. They could still choose among the four recommendations, as in fact they did."

In any case it was natural for me to ask Einstein's advice and help. I did it during one of the afternoons when we worked in his study. We were in the midst of a difficult problem, but the moment I said that I should like to talk about my personal problems Einstein put aside the sheets with the formulae and tried to focus his attention on me. He asked many questions and appeared much concerned with my problem.

"You ought not to go to Poland in these circumstances. We work so well together and we already have important results. I should like to keep you here at least another year. It will not be difficult, I am sure. Wait a minute." He shouted to his secretary in another room:

"Tell me, Miss Dukas, is there any announcement of an institute meeting?"

He discovered that the meeting was two weeks away. He slowly read the announcement:

"There is an item on fellowships. No! It cannot be difficult to secure another fellowship for you. You ought not to worry at all."

I was very happy about our talk. It was not only because Einstein had tried to help me, but the way in which he did it. Going up Mercer Street and then turning toward Nassau, I thought joyfully:

"Life in Princeton is good! The necessity of finding a job, of looking for security, is postponed for a year. In the meantime the paper with Einstein will appear, and this must surely increase my prospects. I shall buy a car and realize the European dream of America. I shall be able to spend the dull Sundays in New York. I shall travel through the great country of which I know so little and devour thousands of miles. I shall travel west and south, visit other universities, make acquaintances and become better known. How clever of me to come to America! I may start to work on nuclear physics. It is difficult to obtain a job here if one is not interested in nuclear physics. There will be plenty of time—a year and a half—to study this field, one which I have neglected up to now. It is much better to stay here a year longer than to take an unimportant job outside Princeton.

America is a serious country. A job here means serious work, often hard work, unlike the University of Lwow. I should have had difficulties with the language, with preparation of lectures, the boredom of elementary teaching, little time for myself. On a fellowship, however, I am completely free. I have only to do research; no other obligation. I shall be able to learn new things. The need for security is removed for another year, but that year will, I am sure, serve to increase my chances."

My thoughts moved over and over the same circular path until I convinced myself that what had happened could only help me.

Two weeks passed, and in our talks we did not mention the subject again. It was not until the day before the meeting that I reminded Einstein, for fear that he might have forgotten it altogether. But it was unnecessary. He had kept the date well in mind and said to me: "Don't worry. I won't forget to go, and I am sure you will get the fellowship—I will fight like a lion for you."

The meeting was scheduled for the next morning. I was working quietly at my desk when at twelve o'clock the telephone rang. I thought:

"It must be Einstein. How considerate of him not to leave me in suspense."

I heard his voice on the telephone:

"I don't want you to be depressed, although the news sounds bad. You did not get a fellowship for next year. I have some ideas about what to do, and you don't need to worry. We shall talk it over this afternoon. But I want you to feel sure that we shall find some way out."

Mechanically I hung up, went to my desk and continued my work where I had left off. Quietly I followed the calculations which were to show whether or not the last step in our reasoning was correct. Half an hour later I went out to lunch. I was afraid to meet anyone and chose a place where the chance would be slight. The food was tasteless. I could not eat. At one o'clock I went home, lay down on the sofa and felt the waves of melancholy wash over me until I was deep in hopelessness. At four o'clock I was to meet Einstein. Still three hours!

"Why didn't I get a fellowship for next year? Doctors Y. and Z., who are no better than I am, got fellowships for the second year, but I did not. In Poland I should have had the simple excuse: anti-Semitism. I always saw that as the reason for all the beatings I ever took in life. But it is much more difficult now to reduce it to this simple cause. Then what can be the reason? Perhaps I have made the unpardonable mistake of over-estimating my place in the scientific world. Is it true that Y. and Z. are no better than I? Perhaps it is only my opinion. Perhaps I am exaggerating the importance of the twenty-five papers which I have written. If one were to eliminate them, science would not change very much. Perhaps I am quite wrong. Perhaps I have over-estimated my own importance without realizing it. Perhaps the institute is not treating me badly after all. I was so sure that I would get the fellowship; was it not conceit and over-confidence? Here I am without a job, without prospects for the future, practically without means, in a strange country, unable to earn any money by teaching physics and mathematics. This I can do well, I am sure. But no one, absolutely no one, is interested in that. I must meet someone. I cannot just stay in this room and think about the same problem in an endless, sterile circle. I must have the courage to face my friends."

To my great comfort I found surprise everywhere. Melbar opened her eyes wide in astonishment.

"But, Ludwik, I don't believe it. Are you sure that you heard correctly over the telephone?"

Robertson's neck got red.

"What? I was sure you would get it." And he comforted me by assurances that he would try to do something about it.

In the afternoon I went to Einstein. I told him of the result of my last calculations as though nothing had happened. It was not a performance on my part. I wanted to get the scientific part of our discussion out of the way and later to talk about my personal problem. Einstein was depressed, much more than I had expected.

"I tried my best. I told them how good you are and that we are doing important scientific work together. But they argued that they don't have enough money and must meet other obligations. There was absolutely nothing personal against you in all this. They all praised you. I don't know how far their arguments are true. I used very strong words which I have never used before. I told them that in my opinion they were doing an unjust thing."

"Did nobody uphold you?"

"No. No one. It was only a meeting of a few of us, but not one of them helped me. But each of them spoke very very nicely about you."

Einstein knew that what I needed at the moment was self-confidence. He stressed many times that I must not take the refusal personally, that everyone thought highly of me. On the other hand he did not want to jump to the conclusion that the decision might have been directed against him. The refusal was as unexpected for him as it was for me, and he could not suggest any explanation.

I sadly explained my situation:

"I cannot go to Poland. Even my friend Loria, who wanted to keep me at the university in Lwow, advised me to stay here. My money is almost spent and will last only to the end of the academic year. I am in a difficult situation. I shall write to Born. Perhaps he can do something for me."

Einstein explained at length that he intended to help me.

"I know that I have a name outside Princeton, and I may be able to raise money for you from some organization. A few years ago I gave a concert for Jewish refugees which brought in six thousand dollars. So I can sometimes raise a little. I don't want you to worry. If nothing else, I can keep you on as my private assistant. I earn more money than I need for myself, and I can give you the equivalent of a fellowship in the institute from my income."

I was touched by this offer but at the same time chagrined. I answered:

"I am very much moved. But you must understand that I cannot accept. I have a definite feeling that it would be bad. If I can't raise money from the institute I want to earn it by my own efforts. Anything else would embarrass me. Perhaps Robertson can find something for me at the university."

We certainly could not have solved the problem at our first meeting, but when I left Einstein I felt much strengthened. Einstein telephoned me the same evening:

"I want you to know that I have already written a very strong letter on your behalf, and if this doesn't work I have another plan. Don't worry! We shall find a way out."

For the next few days I could not concentrate. I gave up my calculations and tried to read detective stories instead. When I apologized to Einstein for the deadlock I had caused in our work he consoled me:

"Take it easy. The world has waited for centuries for the solution of the problem of motion. It can wait two weeks longer."

I tried to return to the calculation of complicated surface integrals, but I could not. It was impossible to do research in this tense state. I gave up my attempts and, lying on the sofa, I thought the same thoughts over and over again.

"How can I earn a modest income here? How can I raise fifteen hundred dollars for next year from the great wealth of America? This is all that I want from the richest country in the world. What am I able to do? I can teach, but nobody wants me to teach. What else can I do? In Poland I had another source of income: writing. But it is gone now. My name is well known in Poland, but it took me three years to establish it. First it was the popular book which I wrote after Halina's death, then my work for the *Gazeta Polska*. But it is different here. How can I write in a foreign language which I speak incorrectly, in which I have a sketchy vocabulary and spell badly and, most important, a language in which I still do not think? To write in Polish and to let someone translate would be to spoil the flow of language by which every piece of writing stands or falls. And besides, how shall I get anything published? I have learned enough about America to know how important it is here to have a name. If I had the same name here as in Poland, my work would be accepted and I should be well paid. But without a name it will be

neither accepted nor paid for. Everything comes with fame and
nothing without it. How is the first step made? It must depend
on chance or on accident. But how can I build my future on the
small probability of an accident? Who will listen to me? Who in
the United States cares what I have to say?"

Then suddenly, from the blue sky, came the idea:

"What about writing a popular book with Einstein?"

The moment this idea occurred to me I knew that it would
solve my financial problem. I knew that a book written with
Einstein, though it might not be a great success, could not fail
completely. I would gain a breathing spell, and the next year
would be assured by the publisher who would, presumably, give
an advance before the book was written.

I was careful to consider all sides of the problem before I
discussed it with Einstein. I knew him well enough to understand that he would never lend his name to ghostwriting. A
book with Einstein's name would really mean a book written
together. But the writing of a book requires time. Had I the
right to take Einstein's time when the only thing in life which
interested him was science?

I knew that if the book was to have any real historical value
I must remain in the background and let Einstein express his
views. It was important that the book should express Einstein's
outlook on science. This I had learned through personal contact
and admired and accepted, with minor differences, but it was
one on which we should certainly be able to compromise. Next
was the problem of the actual labor of writing, the tedious mechanism of popularization which would take a great deal of time.
Here, I was sure, I could do better than Einstein and would be
able to relieve him of most of the work. Some years before, Einstein had written a little book on the theory of relativity which
had been intended to be popular but which had not proved so.
Einstein writes beautifully, rigorously, with a fine touch of
poetry, but always there is the difficulty created by his aloofness.
It is not easy for Einstein to emerge from his inner isolation and
to realize the way in which the ordinary man speaks and thinks.
I was sure that here I could be of great help. I convinced myself

that by writing the book I wouldn't be taking much of Einstein's time nor interfering greatly with his scientific work.

There was another doubt in my mind. Perhaps Einstein would reject the idea of popularization, sharing the view of many scientists who regard popularization as blasphemy, a lowering of the high level of science. I knew that this could not be wholly the case, but Einstein might conceivably have an aversion to a presentation which necessarily avoids the rigorous language of mathematics, omits proofs, quotes only results and simplifies problems and their solutions.

But here I felt justified in trying to convince Einstein since I would be acting in accordance with my own sincere belief. I have always regarded good popularization as a bridge between scientists and society, which reduces the social isolation of scientists; as a deed which, if done well, is of great social importance, raising the level of general thinking, developing criticism, healthy skepticism, combating dogma and creating enthusiasm for the achievements of the human mind.

I expressed these ideas in a sentence which I believe contains some elements of truth even if exaggerated: Hitler would be impossible in a country with a high level of scientific popularization.

I convinced myself that this defeat was the best thing which could have happened to me. To turn defeat into victory, to fight and win, is an essential sign of vital force; it saves one from the dangers of self-satisfaction, routine and security. It was good to feel my brain working again, not merely accepting the situation but looking for a way out. I could not keep my plan to myself. Like a gossipy old woman I talked "in great secret" about it with my friends before going to Einstein. Both Melbar and Robertson thought the idea excellent. We gleefully anticipated the astonishment of some of the institute professors at seeing me still in Princeton the following year, not knowing how and on what money I lived until they suddenly would read the advertisement of our new book. We must not spoil the joke by letting anyone in on the story.

Once more I ordered all the arguments clearly in my mind

before going to Einstein, convinced that I was bringing him a dignified proposition of which I need not be ashamed. Einstein sat comfortably in his armchair, looking at my tense face while I began my well-prepared little speech.

"I should like to talk again about my personal problems. I have thought about them for the last few days. I don't believe that it will be so easy to obtain an outside fellowship to continue my work here, and it would be painful for me to accept money from you. I should like to earn money by more-or-less honest means."

Here Einstein interrupted me:

"How can you? It is not easy. Everything everywhere is crowded. It takes time to find something, and you must solve your problem quickly."

"I believe I have found a way out. I have a plan. It involves your help, but I believe that I won't be making use of you. It all depends on you now."

"What is it?"

I wanted to explain my plan clearly and in logical fashion. We had always discussed everything freely. But, apparently for no reason at all, I could not talk; my well-prepared speech went to pieces, and after a few meaningless phrases, "It is difficult to explain . . . I hope you won't misunderstand me . . ." I gave it up.

Einstein looked at me in utter astonishment. He had never heard me stutter or found me unable to express myself. When I stopped he looked at me for a few seconds, waiting for the rest of the speech.

"For goodness' sake, shoot out what you have to say. I am beginning to be really interested in what it is."

I gathered my courage and began an incoherent explanation, finally making myself clear by repetition, by traversing the subject again and again. I finished with the remark:

"The greatest men of science wrote popular books, books still regarded as classics. Faraday's popular lectures, Maxwell's *Matter and Motion*, the popular writings of Helmholtz and Boltzmann still make exciting reading."

Einstein looked at me silently, stroked his mustache with his finger and then said quietly:

"This is not at all a stupid idea. Not stupid at all." Then he got up, stretched out his hand to me and said: "We shall do it." The fate of our book was sealed. The book was born.

We had decided to write the book but we had no idea what it would be about. I suggested the theory of relativity; to explain the principal ideas in so simple a way that any intelligent reader, without knowledge of mathematics, would be able to grasp them. Einstein was not enthusiastic about this suggestion, claiming that too many popular books about the theory of relativity had already appeared.

Without knowing what we should write about, without knowing when the book would be ready, I decided to talk to a publisher. Thus I entered a world of new experience, a fantastic world from the point of view of a scientist—that of the publishing business.

My first visit was to Simon and Schuster, whom one of my friends in Princeton recommended. Neither Simon nor Schuster was in the office on the occasion of my first visit. I talked to the charming and vital Maria Leiper, an associate editor. She was immediately enthusiastic. What an honor, how happy they would be, etc., etc. Of course I lied. I said that we had a part of the book ready to submit and that we were looking for a publisher. In answer to a direct question I said that one third of the book was written but not in final form. I took comfort by telling myself that publishers are used to lying authors and are not inclined to take their statements too seriously even when they are telling the truth. My lies made it possible for us to discuss the nonexistent manuscript concretely. From my description Miss Leiper formed a picture of a book, its drawings, jacket and price, although I still did not have the slightest idea of its contents. But it was all right. Publishers always know better than the author, even as to what the book should contain.

New York had always depressed me. The height of the skyscrapers, the overwhelming wealth of the city, the restlessness of the people, the tremendous energy wasted in the subways by gum-chewing mouths, repelled me. But now, when leaving the publishers' office, I saw the city in a new light. It was the most wonderful city in the world! Success radiated from these noisy streets. The town would be mine! I felt like one of the millions of restless New Yorkers looking for the treasure lying about on the pavement and which fell to those who knew the magic word. I knew from the response of the publishers that my financial difficulties were solved for the next year and that there was a possibility of a success beyond my first modest dreams.

I had practically sold the book. From an ethical point of view I was a swindler selling something which did not exist. In order to attain a semblance of virtue I hurried back to Princeton to talk further about the book and thus make it more real.

Once Einstein absorbed the plan he was most enthusiastic about it. Never once while writing the book did he take his task lightly. With every day he found the work more attractive, repeating:

"This was a splendid idea of yours."

We discussed, revised, discussed and revised until the book was in its final form. Then suddenly Einstein lost all interest. His interest lasted exactly as long as our work lasted. It ended the moment our work was finished.

The original idea for the material of the book came from Einstein. His intention was to write a popular book containing the principal ideas of physics in their logical development. His point was that there are amazingly few fundamental ideas in physics and they can all be represented in words. As he said:

"No scientist thinks in formulae."

Before a physicist begins to calculate he must have some picture or some pattern of reasoning in his mind which can, in most cases, be formulated in simple words. Calculations and formulae are the next step. Thus it would be our aim to represent the most fundamental thoughts in physics in their proper places and perspective. Einstein was sure that it would be possible to do this in one volume, covering all branches of physics.

When we talked of this plan for the first time Einstein was ill. He lay in bed without shirt or pajamas, with *Don Quixote* on his night table. It is the book which he enjoys most and likes to read for relaxation. I sat near him on a chair. The thought of our book excited him; he sat up in bed, exposing his big naked torso, and said:

"It is a drama, a drama of ideas. It ought to be absorbing and highly interesting for everyone who likes science."

I understood his plan perfectly, and after this one brief talk I had a clear picture of our book, a picture which I tried to work out for myself in more detail. The months of collaboration brought a splendid understanding between us, one in which a few words stood for pages of reasoning. We were able to communicate with each other by means of a sort of shorthand speech.

I liked Einstein's plan. It was a difficult theme, and the choice of ideas, the selection of those to be stressed, the establishing of links between them, are all determined by the physicist's personal taste. But for this very reason it ought to be important to learn Einstein's view on this matter. No one else has thought out all the basic problems in physics as deeply as has Einstein. No other living physicist is as qualified to write on these subjects with such authority. Whether or not the book would be a financial success became less and less essential to me. An important book would appear, and solely because I had come to Princeton. The thought made me proud and happy.

The whole idea of the book changed. It was supposed to have been a simple popular exposition. But now it must be more. It must be a book from which I myself and other physicists would be able to learn something; not concrete knowledge of facts, but a new point of view, ideas placed in an original perspective. The book would and must be a scholarly work, but at the same time it must be written simply; no previous knowledge, but a rather high intellectual level, should be assumed on the part of the reader.

The next time we talked about the form of the book we understood that both the writing and the presentation must be simple. We both felt alike about the aim of popularization and despised the attitude of the scientific writer who tries to play on the reader's emotions. To hold the attention of the poor reader such writers often attempt to be witty, using jokes so irrelevant that the joke remains in the reader's mind and not the reason for which it was introduced. They jump suddenly from physics to metaphysics, persuading the reader to turn the pages even though he does not understand. In such books the universe is pictured as immense, the galaxies as millions of light-years away, and the human being is terrified by this cruel and empty cosmos. The atom appears fantastically small when compared to the universe. And above all there is religion; God is a great mathematician, the greatness of science reflecting only the glory of God. The aim is to stress the results which seem to contradict the common sense of the ordinary man and to persuade him how clever and intellectual the scientists are.

But beauty can be found everywhere: in the laws governing the falling stone and in the motion of the galaxies. Mystery, being everywhere, is therefore nowhere. Emotion can be created by the effort of understanding, by the painful process of comprehending more fully and more deeply.

But to effect this kind of popularization means to forbear startling and spectacular effects. It means straightforward writing; it means an attempt to convince the reader that science is merely a highly developed common sense and that its aim is to order the world of our experience and to form a consistent picture of reality.

The problem of language presented another difficulty. I had never written anything in English, with the exception of private or official letters which were always corrected by someone. I discussed the book with Einstein in German, thought in Polish and had to write in English. And I knew that a translation would not do.

I tried my best and slowly put the phrases on a paper in incorrect, badly spelled English. I rewrote and corrected until I found myself on the outskirts of a very restricted knowledge of the English language. I had good friends in Princeton and most devoted and loyal aid from three of them who, in succession, helped me, discussed my obstinately repeated mistakes and taught me to write English. At the end of the book I scarcely realized that I was not writing in my native language.

Einstein and I systematized our work so that the book and the paper on the problem of motion would not interfere with each other. We discussed the book only at two-week intervals. I read my portion to Einstein, who invariably repeated before each session:

"Read it slowly, so that I can understand each word."

He listened attentively and made his remarks. As a rule there were only a few changes. We discussed them carefully, trying to reach an agreement. Our collaboration went most smoothly. Einstein made it a point not to interfere with the form of presentation, repeating:

"I don't care how you say it. You know better. But this idea must come into the book."

I often objected on the ground that the idea was too complicated and that it would be better to omit it rather than to discourage the reader. We always reached some kind of a compromise, and the state of the reader's mind was our greatest concern. After having read and talked over the portion just written we would go for a walk, discussing the next section, always trying to harmonize any differing opinions. Never during our collaboration did Einstein act authoritatively. As the work progressed we understood each other better and more quickly; our meetings concerning the book were shorter and differences of opinion more easily settled. Each day I wrote about one thousand words. Step by step the book began to take shape.

My first year in Princeton was nearing its end. During the vacation months Princeton is one of the most uncomfortable places in the world. The town is dead. The few remaining inhabitants, moving languidly over the burning pavement, repeat over and over in pathetic voices that the humidity is to blame. I sweated and drank water, drank more water and sweated, got up at five to write my thousand words before the burning sun made a furnace of the town. Einstein was away. We had discussed the last two chapters before he left, and I went twice to his summer place to make the final changes in the manuscript.

Except for the title, the book was finished. Just before Labor Day I sent the manuscript to the publishers and went off on a two weeks' vacation. We made up our minds to refuse to make any changes the publishers might suggest. This stubborn attitude on my part was due to my belief, probably exaggerated, in the importance of our work. The project had begun in defeat, as a way out of my financial difficulties. But once we became absorbed in the book neither of us made the slightest attempt to cater to the public in order to increase the sales. Lucidity was our only demand, and I never allowed myself to cheapen the book by anything which was effective but unnecessary. I felt always the weight of Einstein's name and my responsibility in the collaboration. When I bothered Einstein later about details of representation I explained to him:

"I should have been much freer and less careful if I had been the only author. But I cannot forget that your name will appear on it."

Einstein laughed his loud laugh and replied:

"You don't need to be so careful about this. There are incorrect papers under my name too."

It was a relief to be rid of the manuscript and to return to the scientific problem of motion. While the book had not seriously interrupted our scientific work, I now felt that I could peacefully check our previous reasonings and calculations.

It was autumn 1937 when I went up to New York to make the final arrangements with my publishers. Their behavior was above reproach. At this time Mr Simon read aloud the reports of several readers to me, all of which stated the importance of the book, foresaw a steady but perhaps not spectacular sale and stressed that not a word in the manuscript ought to be changed.

I found the publishers' office an attractively crazy place where everyone is interested in books and no one reads them. The designer of the books, who later became my good friend, made it a point never to be impressed by anything and always to predict defeat. He was interested only in the physical aspect of the book. At least he had the excuse that he did not need to read books inasmuch as he collected them, designed them and was himself a successful author. For him a book by Einstein was simply a problem in design, a book attractive enough to be chosen the best-designed book of the month, as it later was. He wrote an excellent blurb without having read the book. For the advertising manager the object was to select some vivid phrases from the reviews and to do a proper job of advertising without looking into the book. A publishing house is an exciting place where books are born, pushed or butchered. The atmosphere is fantastic. Here books are produced, published, sold, converted into money—sometimes much money—but they are not read. They are absorbed by a sort of osmosis, through the senses of touch and smell, through letters from the readers, so that everyone in the business knows and can talk for hours about any book published without having read it in the way ordinary people do, by opening it and turning the pages. Everything that happens around the book—the designing, the sales, advertisements, reviews—is important; the publishers and their staff love it. But they detest books. To them authors are queer birds, pathological, oversensitive, conceited, who must be handled with the care given an uncooked and slightly unreliable egg. It was an amazing show, and I spent some lovely hours in that crazy place.

When my second year in Princeton began my self-confidence had increased. I was secure for the next year or two, and I enjoyed the sight of the astonished faces I encountered in Princeton. Einstein reported to me the following conversation: one of the influential members of the institute asked Einstein what I was doing in Princeton. Evidently Einstein was still very angry about my previous rebuff, for he answered diplomatically:

"Since Infeld did not get the fellowship, I did something to secure him for this year so that we could finish our work."

"How did you do it?"

"I am not at liberty to tell you."

Any success, even a moderate one, influences our relations with other people. I now felt quite differently when talking with the important members of the institute. The old strain of keeping on guard, the care in choosing words, the fear of making the wrong impression, the sickening analyses of other people's behavior toward me and my behavior toward them were gone. I gained a freedom of expression which in turn allowed me to be more natural, to be kinder and more sympathetic toward others.

A few months before, during the days when I worried about my future, I had thought: "Very likely Professor X. is responsible for my not getting the fellowship." I suspected that for some unknown reason the idea of damaging my future was one of the basic aims of his life. Once having suspected that Professor X. deliberately harmed me, I saw my suspicion confirmed by all his words and actions and watched for gossip from other people for confirmation of my attitude. An inferiority complex nourished by defeat leads directly to a persecution mania. But a modicum of success changed my whole outlook. The same Mr X. appeared as a basically decent man who had acted according to his best convictions and was only doing what he thought was best. The pressure was lifted, life became good again, people seemed kind and human bonds precious.

After our suggestions for a title were rejected by the publishers, and after the publishers' suggestions were rejected by us, we finally agreed on *The Evolution of Physics*. When the advance copies arrived I took them to Einstein. He was completely disinterested, and to this day he scarcely knows what the book looks like. External things are trivial for Einstein and he can hardly imagine that appearance can make the slightest difference to the book. The publishers asked me many times: "How did Einstein like the finished copy?" To which I always answered: "He liked it very much," not wanting to upset them by telling them that Einstein had not even opened the book. Once a work is finished his interest in it ceases. The same applies to the reprints of his scientific papers. Later he had to autograph so many copies of our book that automatically, when he saw a blue jacket, he groped for his fountain pen.

The day before the book appeared a reporter from an important newspaper telephoned Einstein from New York at eleven at night, asking him whether he would like to say a few words about the book. To this Einstein replied:

"What I could say about the book is in the book."

Reporters came to Princeton and interviewed me. *Time* decided to run an article about *The Evolution of Physics* with Einstein's picture on the front cover.

When the publicity about the book began I became important, for I had done something which everyone in Fine Hall detests and desires. I had made the general public conscious of my existence. I no longer sat alone at teatime in the commons. Influential professors who had hardly been aware of my existence joined me now with bright smiles, half ironical and half benevolent. Questions were shot at me:

"Have you seen the full-page advertisement of your book in today's New York *Times?*"

"How is it going? Have they sold ten thousand yet?"

"I bought your book. The chapter about relativity is excellent. I liked the one about quanta less."

"Who really put the book into English?"

"I should like to give a copy to my brother. Could you ask Einstein to autograph it?"

"Did you see your book on the best-seller list? It jumped above *How to Win Friends and Influence People*. Einstein has beaten Dale Carnegie this week."

"It would be interesting to know who buys the book. It is not easy at all, though very well written. I wonder how much of the success is due to Einstein's name."

I was invited to dinner where before I had been invited to tea; I was invited to tea where I had not been invited at all.

Only Robertson remained the same and teased me, saying that although he had reviewed our book he had not and would not read it.

I began to savor the taste of fame. It is a tingling, pleasant and faintly bitter taste. One can get used to it quickly and miss it when it fades. I thought:

"I am exactly the same man I was three months ago, and I have not made any important progress in research since then. A year ago I would have accepted any job anywhere. Now I see my picture in the papers and even here in Princeton, where I should have imagined only scientific achievement matters, where I thought writing a popular book would be regarded rather as a disgrace, even here people have changed their attitude and behave differently. I must watch myself carefully and not make the same blunder that others do; I must not overestimate external and unimportant events; I must be on my guard and learn from Einstein that the only change which success should bring is increased kindness."

Mine was a perfect example of reflected glory. To the end of my life I am stamped as the "collaborator of Einstein." The book —to the man on the street—was written by "Einstein and somebody else whose name I don't remember." Even now I am spoken of as the "man who worked with Einstein."

However, I certainly had my compensations. I had found a way out of a hopeless situation; I no longer needed to look for a job and I had increased my freedom. The book was much more successful than we had expected, and many editions were printed. I know that physicists read the book. But it is and will remain a mystery to me who bought the many copies that were sold. I asked my friend the designer at Simon and Schuster:

"Who does buy books?"

His reply was:

"I've worked for ten years in the publishing business in order to find an answer to that question and I still don't know."

Letters began to pour in from all over the world. Only a very few were sincere and sensible. Most of them were from people who had not even seen the book and who, in their "cosmic illness," devised their own personal theories of the universe; or else they were letters full of abuse and indignation.

There were also pleasant letters. One was from a clerk in Chicago. It contained a simple question. Was I the same Infeld who twenty-two years before was in the Austrian army in the barracks near Cracow? It was written by the good and kind corporal who had brought me passes and put a check mark beside my name during military inspection.

Another letter came from Joseph. We had lost each other after Hitler had come to power. From a review of the book he had guessed that I was in Princeton, and I was happy to learn that my friend was safe outside Germany.

In its first three months our book oscillated from the bottom to the top of the best-seller list and down again, until it no longer appeared and its place was taken by others. Letters came less and less often; the whole fanfare faded slowly away.

But my increased self-confidence remained. I knew then that I need not worry about my daily bread.

X

MY SECOND YEAR in Princeton drew to a close and I had done nothing toward finding a job. I intended staying in Princeton for a third year. The sales of *The Evolution of Physics* had made this possible. I planned to relinquish my academic career and earn my living by writing. At the same time I guessed from Robertson's vague remarks that he was exchanging letters with several universities which were considering me. The first offer came from Canada. Professor Synge, head of the department of applied mathematics in the University of Toronto, offered me a lectureship, a temporary appointment for one year. He expressed the hope that later it might become a permanent arrangement. My duties would be to lecture six or eight hours a week, mostly on advanced subjects in mathematical physics. I knew that the University of Toronto had one of the best undergraduate mathematics departments on this continent, a good graduate school and that the head of the department, who wrote to me, was well known among scientists. His offer reminded me of the letter which I had received ten years before from Loria, through which my university career had begun. Even the circumstances were similar.

A lectureship is an excellent appointment for someone who has recently taken his Ph.D. and is at the beginning of his scientific career. But I was well aware of the depreciation of scientific values, measured in U.S. or Canadian dollars, caused by the happenings in Europe. About this time I met a splendid mathematical physicist from Germany who had been invited to America for a short lecture tour. Although he himself was "Aryan" he had had to leave his country because his wife was Jewish. His papers were considered classics in theoretical physics, and in Germany he had been head of a department of a great university. At a moment's notice, he told me, he would be glad to accept a position, even one as modest as that which was offered to me. The law of supply and demand regulated the scientific market, and the fact remained that many distinguished scientists could be bought for very little money. Famous mathematicians from Germany and Poland accepted instructorships in colleges which hardly deserved the name of university. Of course I knew that there was another side to this problem. When I obtain a position in Canada I occupy a position to which a Canadian could have been admitted. The argument often used is that the ingress of foreign scientists diminishes the chances of native scientists. But there is again a third side to this problem. By raising the level of the universities new centers of scholarship may be created, new needs may arise, in the long run increasing the number of openings and, what is more important, creating an atmosphere in which science in America will develop more richly. For a long time to come America can be justly proud of having created new opportunities for scientists expelled from Europe and of saving their contributions to science from being lost forever. Anyone who argues that the positions were selfishly offered, that no one profits as much as the American universities, does not understand the simple truth that all progress springs from a combination of egoism and good, clear, farsighted thinking.

I heard of a scientist, the best in his specialized field, who had two appointments: he spent half the academic year in Germany and half in America. When Hitler came to power the professor resigned from his position in Germany. He finished his letter of resignation ironically by expressing the hope that the German minister of education might succeed in raising the level of German universities during his whole future life as much as he had raised the level of American universities during the first three months of his term of office.

Einstein, Robertson and everyone else with whom I discussed it strongly urged me to accept the offer from the University of Toronto, although Einstein remarked many times how much he regretted that we should have to interrupt our collaboration. Robertson's argument was: "It is a job, and a job is a job. You have to start again as if you were a young Ph.D. But the difference is that once you have started you can go ahead quickly." I decided to go to Toronto and accept a position very similar to the one in which I had begun my scientific career in Lwow ten years before.

Many times while writing these pages I have tried to answer the question: "Am I honest with myself?" I have not written the whole truth, but everything that I have written is *my* truth. There have been events in my life of which I am still unable to write. The most important gap in my story concerns the vacation period after the second year in Princeton. In this interval I lived through the dramatic epilogue of a relationship which I have consistently omitted from my book.

For some days before I went to Toronto I stayed in New York and attended a meeting of the American Mathematical Society. There I met Helen; it was ten years after I had first seen Halina at the Wilno meeting and six years after her death. When I met Helen I was unhappy and still distressed by the delayed aftereffects of my past tragedy. Helen helped me to understand the relation between my present state and my past experience, to see cause-and-effect connections between events which seemed to me unrelated. I told her the story of Halina's tragic death; I told her of the thoughts, the scenes, the details which I had suppressed for years. The healing process was quick and thorough; I was restored by Helen's sympathy, love and understanding.

I had to leave New York ten days after we met, but we both knew that the bonds between us were to last our lifetime. For the next seven months our work kept us five hundred miles apart.

In Toronto I felt cut adrift. Once, and sometimes twice, a month I went to New York for the week end to see Helen and returned just in time for my Tuesday morning lecture. My life was concentrated on these brief visits, and the intervals between them were filled with expensive telephone calls and daily letters. I am astonished now that in spite of my longings and the arduous preparation of lectures I could still do scientific work.

Toronto is a curious mixture of a town in the United States and one in England. Externally it is like the United States. The same drugstores with milk shakes and sodas, the same cars, tourist homes, cabins, advertisements. But it is different if one looks beneath the surface. The silence, the reserve, the slow tempo, are those of an English town. In the evenings streets are empty, and their deadly silence is oppressive. On Saturday afternoons a new spark of life frightens the town, only to die out completely on Sundays. For one day the town becomes lifeless, only to show a slow, scarcely perceptible pulse on Monday. It must be good to die in Toronto. The transition between life and death would be continuous, painless and scarcely noticeable in this silent town. I dreaded the Sundays and prayed to God that if he chose for me to die in Toronto he would let it be on a Saturday afternoon to save me from one more Toronto Sunday.

Slowly, beneath the city's armor of tranquillity, I found friendly, helpful and loyal people. Toronto is a good place for the evening of life, for raising a family, for relaxation and quietly watching the days pass. It seems fantastic that anything irregular can happen to anyone in Toronto.

The department in which I worked is in a small, ill-kept building. It was strange to climb the nearly disintegrating stairs, through simply furnished rooms, after having worked in Fine Hall's luxurious décor. I lectured to small classes, to graduate and fourth-year students, who took notes diligently and silently, maintaining their serious faces in spite of all my attempts to make them laugh. It was months before I succeeded in producing slight grins on their faces, but before the year finished they needed little provocation to burst into loud laughter. The students in Toronto are a serious crowd, trying to keep up, by hard work, with the high standards of the university.

The head of our small department and my few colleagues had not the slightest idea of university intrigue. Perhaps they knew, but they did not show it by their behavior. Their decency was almost painful. I was surrounded by such a sea of politeness, by such perfectly gentlemanly behavior, that I felt a desire to say to them: "Try to be a little gauche, it will be more amusing; do it for me, please, to make me feel more at home; talk more, talk louder, swear sometimes; believe me, it will do you a lot of good."

Years ago when I returned from Cambridge to Poland I tried to play the role of an English gentleman. For a few days I talked politely, slowly, in a low tone, keeping my boisterousness firmly inside. God knows how much effort it cost me and how silly I looked. Now I have outgrown this stage and given up all pretending. In Toronto I heard my loud voice amplified by the persistent silence; my gestures seemed to me spectacular against this calm background. The dignified, tranquil way in which everything was accepted redoubled my efforts to shatter this unresisting poise. I felt that I must be shocking to my surroundings, but my surroundings refused to show that they were shocked.

Near the end of the academic year the university offered me a permanent appointment, a professorship with a modest salary. For twenty years my ambitions, desires, plans, had been concentrated upon becoming a university professor. When I finally obtained the offer I was neither very astonished nor very much impressed. Perhaps I had outgrown the dreams of my youth before they had come true, or perhaps I regretted that at last everything seemed to be settled for the rest of my days. Of course I accepted the offer.

The seven months in Toronto seemed to me like a distinct border line between two lives: my past—with its adventures, its struggles and disappointments—and the peaceful future. I wanted to accelerate the flow of time, to be a few months older, to cross this border line between my two lives quickly. Literally I counted months, then weeks, then days, then hours, until the lectures would end, until I could marry Helen. I finished my lectures on the tenth of April, and the university calmly received my announcement that I should like to go away and wanted the examination papers mailed to me. Although the registrar said: "This is a little unusual," it was done as a special favor, and I had a five months' vacation period before me.

A judge in New Jersey pronounced us husband and wife and I moved into Helen's apartment. I am ashamed to write down this trivial sentence. It disparages our emotions, the experience of adjustment, the excitement of starting a new life together, all too fresh and vivid in my memory to be expressed in any but words that touch only the most external surface of events. But there was one personal aspect of our relationship which helped me to understand myself and, I feel, ought to have a place here.

The difference between the environments in which Helen and I were brought up was very great, indeed so great that it would be difficult to imagine it any greater. The worlds of our childhood were separated by thousands of miles in space and by centuries in time. The atmosphere of Helen's home was in the best liberal American tradition, cultivated by a family whose roots had grown for generations in this country. Helen was taught to detest social and racial prejudices and to look upon anti-Semitism as an old weapon of reaction. Her background symbolized for me the world to which I had always wanted to escape from the isolated island of my childhood. Did the attraction to this world influence my emotions? I would never have formulated this question for myself. I would never have looked for a connection between my love and my experiences of over thirty years before. It was Helen who made this relation clear to me, and I do now believe in its existence. Like a successful scientific theory, it also explains other changes through which I went after I met Helen.

In the days past I always suffered whenever I was reminded that I was a Jew and was happy for occasions, such as in Cambridge, when I was allowed to forget it. I had suppressed this though I felt embarrassed, ashamed or provocative and arrogant whenever this problem arose. Only after I met Helen did I discover that I could think of the elapsed years freely and calmly. I gained the courage to look back, and I learned to understand the influence of past events upon my life because I was nearer to the desired outside world than at any time before.

How is it possible that it took me so many years to see things which, once grasped, seem so unmistakably clear and simple? How is it possible that I needed outside help to place them in the right perspective?

As a physicist once remarked to me, everything seems trivial after it is well understood. A problem which appeared difficult yesterday may appear trivial today if we have, in the meantime, found that it has a simple solution. The same may be true when we try to understand deeds and experiences, both our own and others'. They may seem strange and unrelated until we are able to deduce them from simple traits of character. The meaning and the motives for our actions are easily covered by self-deceit. For many it provides the glasses which deform reality and diminish the chasm between desire and fulfillment. The self-deceit grows most easily if this chasm is deep and wide. One of the very few things which I believe I have achieved and which I value is that I have decreased my self-deceit as the years of my life have increased.

Once when Helen was busy with examinations and meetings I decided to go to Princeton for a day, to see Einstein and Robertson and to look once more upon the place of my last struggle and of the greatest scientific experience of my life.

During the months in Toronto I had worked on the problem of generalizing the equations of motion, the same problem on which a year before I had worked with Einstein. Often in theoretical physics, once the first break is made and a new difficulty conquered, one finds that some of the assumptions are unnecessary, that the mathematical structure may be formulated more precisely and more generally. I tried to eliminate the special assumptions of "co-ordinate conditions," to find the equations of motion not in a *special* co-ordinate system, as we had done before, but in an *arbitrary* co-ordinate system more truly representing the spirit of general relativity theory. I had sent Einstein the first draft of my manuscript. He liked the general idea and made two remarks which put the problem in a new perspective. He pointed out that not only can the equations of motion be formulated in this more general way, but we may also indicate the method of their solution by using the same approximation procedure which we had employed before. Secondly he suggested a simplification, skillfully twisting and changing my argument, looking at the problem from a fresh and original angle. I wanted to discuss all these things with him.

After a year's absence I entered the house on Mercer Street. It took me a little time before I felt as free in his study as I had when we worked on "The Problem of Motion" and on *The Evolution of Physics*. We discussed the new results achieved through our collaboration by correspondence, and Einstein suggested that we publish them together. It was decided that I should write the manuscript and send it to Einstein for his approval. I was glad that our collaboration had been prolonged in spite of the geographical distance. Then Einstein told me of his new attempts at building a unitary field theory, of his disappointments and hopes, repeating several times:

"I am sorry that you are not in Princeton. We understand each other so well. It was fun to work together."

I felt depressed during this visit. One of the reasons was the realization that the tremendously inspiring experience of working with Einstein would have to end. The other reason was that in our talks we drifted toward social problems, and to Einstein everything seemed darker than ever before. I was affected by his pessimism. He thought that the future of Europe had been determined by the events in Madrid and Munich and that "fate marches on." Never before had he found the political situation so hopeless and chaos so near.

It was like one of the many talks we had had when we worked together. Later, on my way to Fine Hall, I crossed the campus full of sun and spring flowers. The well-known scenery made me realize more vividly the changes which the last year had brought to me. I no longer felt the old tension and the constant fear of the days to come. I could look now with assurance and calmness into the future. I saw in it a happy family life in a town splendidly suited for it. Everything seemed to be perfectly settled for the rest of our days on this earth. After a while we would have a child, then another. I should write papers, books, get a raise in salary from time to time, live in the best part of the world, in the warm atmosphere of a home created by a loving and understanding wife.

Excitement would no longer come from external events but from our common work, from thoughts, books, emotions and friendships. Helen would help me to destroy the restlessness accumulated in the years of struggle. I would enjoy the small details of peaceful, quiet days and await calmly the hour of my death.

But will I be able to destroy the restlessness which up to now has grown from the years of my childhood? Perhaps I will miss the atmosphere of fight and struggle which I have breathed for so many years. Have I really crossed all the bridges leading to the outside world from the island on which I was born? Perhaps I will be forced to retreat and to start my wanderings again if darkness and hate spread over the world.

Passively to await the future means to approve of the world of today and to share the responsibility for its fate. Where is my place in this world?

XI

IT HAS BEEN a long journey from Cracow's ghetto, from
the banks of the Vistula River, to the University of Toronto.
There is still a name in my memory which has always remained
a symbol of lost hopes: it is *Konin*. It is the name of a small
Polish town where I lived for two years. In this town were
several thousand inhabitants, mostly Jews. There were no pavements, no library, no movie, not one house with a bathroom.
While I was there my world was divided into two parts: isolated
Konin in which, I thought, I should probably die, and the rest
of the world which I should never see. Each evening I saw the
same two prostitutes walking for hours round and round the
square market, where twice a week peasants brought their
produce. I knew each face in the town and I was known to
everyone. Many doffed their hats ceremoniously when they
saw me, because already, at the age of twenty-five, I was the
principal of their Jewish high school.

The school was crowded into a tiny red brick building. It
comprised a little more than a hundred students. My colleagues
were young and even less experienced than I. Wherever we
went we encountered each other. We got on each others' nerves,
and wild quarrels, accusations and counteraccusations broke out
between us for the most trivial reasons. Our salaries were nearly
worthless a week after we received them. They were measured
in thousands of Polish marks at the beginning of the school year
and millions at the end. The school was in constant financial
difficulty and each month, we thought, would be its last. I lectured to the parents on the importance of learning and concluded each speech with an appeal for money, lest the teachers
die of starvation. When I returned home I could not bear to

look at my scientific books, collected during years of study. I did not believe that I would ever open one of them again in my life.

Near the school building, on the shores of the river Warta, there was a much larger, unfinished structure of red brick. The story of these unfinished walls grew to be a legend of Konin, reflecting the sorrowful history of Polish Jewry. This is the story as I heard it:

Sometime before I went to Konin the town had been rich. It was rich because there was no larger town within a radius of thirty miles, because there was no railroad in the vicinity, no cars and no good roads. Konin was the center of a small, isolated section of Poland. The Jewish citizens decided to do what citizens of other towns had done after the creation of free Poland: to build a high school for their children. They were proud of their project and wanted their new school building to dominate Konin's sky line. They planned, therefore, to build a three-story building with a small tower (for astronomy, so they said). The money was raised, a picturesque place on the river Warta chosen, a Gentile constructor hired, and the walls began to grow. But the non-Jewish minority could not bear the thought that the tallest building of a Polish town should be a Jewish school. They put their heads together and decided that something must be done. By picturing the danger of Jewish domination, and by bribery, they persuaded the builder to erect one of the walls without the proper foundation, nearly straight from the ground.

But in those days God still took some interest in the fate of the children of Israel and prevented the castrophe in His strange and miraculous ways. First the Polish government built a railroad near the town. Konin's decay began, and the poverty of its citizens increased. Subscriptions for the school building ceased to flow in, and the four naked walls stood idle for months, one of them without foundations. The children were still being taught in the tiny rented building. Instead of collapsing suddenly, the wall began to crack and curve. The plot was discovered. Many times I heard the story of the meeting between the trustees of the school and the builder. One of the trustees whom I knew, a fat, bearded merchant, had a talent for making this scene vivid with inimitable gestures:

"He sat on his chair like me now. So he sat." Here he put his head humbly down, imitating the builder. "So I get up suddenly and say to him: 'You kill Jewish children, you'll be hanged. New law in Warsaw now.' So he got white as this paper here, not so, even whiter. He fell down." Here the storyteller fell on his knees, raised his hands and gave a loud cry: "Oh, I have a wife and children, have pity, have pity on me. I will do what you tell me, I will build a new good wall. Have pity! Don't tell the police."

The conditions of the peace were harsh. The builder had to construct a new wall and reduce the price of the building.

All that happened before I came to Konin. When I left Konin for Warsaw two years later the school building was still un-finished.

After Warsaw there was Lwow and then the Rockefeller fellowship to Cambridge.

One winter evening I sat on the top of a London bus going toward Piccadilly Circus. I was on my way to the semiannual supper given by the Royal Society to all authors who had con-tributed to the *Proceedings of the Royal Society* during the previous half year. In accordance with the instructions on the invitation I wore tails, and I had been snobbish enough to order them from one of the most expensive tailors in London. I was deep in thought when suddenly I heard someone behind me talking Polish to his neighbor. I turned my head and saw Mr M. from Konin, one of the more intelligent members of the Board of Trustees. He was very happy to meet me. He had read in a Polish newspaper that I was in Cambridge. Konin, he as-sured me, was very proud of me and took a great interest in my career. So delighted was he to see me that he paid two extra pennies and accompanied me all the way to Piccadilly Circus.

"I couldn't bear it any longer in Konin," he said. "I took all my savings and ran away. I went to Paris, from there to New

York. Now I am in London, and in a week I shall be back in Konin."

"How is the school?"

"It doesn't exist any more. It was liquidated two years after you left. All the wealthy people have left Konin. The town becomes poorer with every passing day."

"What happened to the new building?"

"The government wants to use it for a prison. It is still unfinished."

I remembered reading somewhere that "who builds schools destroys prisons." It is different in Konin. Who builds schools builds prisons.

He repeated several times: "Isn't the world small after all!" which was exactly what I expected him to say. Then we shook hands warmly and, conscious of my splendid tails, I entered the building of the Royal Society.

This story of Konin's school and the incidents in London have become for me a symbol. In condensed form they reflect the misery of those whom I have left behind and my fortunate escape.

Even in America years later I still received letters from my old pupils in Konin, begging me to help them emigrate to this country. They were among the many letters I received from Europe from people whom I knew and from people of whom I had never heard; all of them invariably sent by registered mail, full of pathos, sent by men and women who, for the price of a postage stamp, bought hope for a few weeks, waiting for an answer which in most cases buried this hope.

A source of light radiates energy. An obstacle placed in the way of light rays creates shadows. Light rays seem to travel along straight lines. But we now know that this picture is too crude, too primitive to describe the phenomena of nature. The wave theory of light reveals that there are no distinct shadows if the wave length is comparable to the dimensions of the obstacles.

Ocean waves impinging upon a ship appear on its other side.

They bend and curve, revealing the phenomenon of diffraction. According to Huygens' principle, each point reached by a wave becomes the center of a new, though weaker wave spreading in all directions.

This physical law is obeyed by the ocean waves, by the electromagnetic waves of light and also by the waves of hate. Their tendency is to penetrate everywhere. They will not reach our shores if we can build obstacles and walls much greater than the length of the impinging waves. The length of the wave of hate is great, and so the protecting wall must be still greater.

Is there a harbor, is there a haven for me and thousands of others? Am I now safe behind a wall which is large and strong enough to resist the impact of hate?

I do not know. I do not know whether my experiences in Poland represent a pattern or an isolated case. In the last years I spent in my native country I saw Poland's fear of the increasing might of the German Reich. The protective wall against it was built by treaties and by the fourth strongest army in the world. But the waves of hate penetrated easily through this wall. The waves were much weaker than in Germany, but their amplitude increased constantly. We did not have inscriptions, *Jews not wanted*, but there were notices in hotels: *For Gentiles only*. Jews and radicals were not put in concentration camps as in Germany, but, much less brutally than in Germany, they were removed from the Polish cultural and economic life. There was no law, as in Germany, destroying freedom of speech, but those who talked too loudly against the administration were put in prison. There was no law in Poland that a Jew could not be a member of the University Assistants Society, but the Assistants Society in Lwow expelled me by its own decision. There were no laws that Jews must not study at the university, but the student hoodlums knew how to replace law by force.

Is this a pattern which will be repeated? I have already heard in America the familiar words: "Anti-Semitism is increasing in America." I have read inscriptions: *Restricted*. The waves of hate are incomparably weaker here than in Poland. They were weaker in Poland than in Germany. But am I witnessing here the beginning of a process of which anti-Semitism is an external sign, or are these only isolated insignificant cases and not characteristic of the greatest nation in the world?

I do not know the answer. No one knows it. But to a small degree the answer depends on me. What I do in life contributes one small component to those whose summation gives the great resulting force upon which the answer depends. And when I write this book I do it in the conceited belief that I am putting a brick into the wall which may throw off the waves of hate.

Where is my place in this struggle?

I belong to the great family of scientists. Each of us knows that curious state of excitement during which nothing in life seems important but the problem on which we are working. The whole world becomes unreal and all our thoughts spin madly around the subjects of research. To the outsider we may look like idle creatures, lying comfortably about, but we well know that it is an exacting and tiring task that we perform. We may seem ridiculous when we fill sheets of paper with formulae and equations or when we use a strange language in our discussions, composed of words understandable only to the initiated. We may look for weeks or months or years for the right way to prove a theorem or perform an experiment, trying different pathways, wandering through darkness, knowing all the time that there must be a broad and comfortable highway leading to our goal. But man has little chance of finding it. We experience the ecstasy of discovery in very rare moments, divided from each other by long intervals of doubt, of painful and attractive research.

We know these emotions so well that we hardly ever talk about them. And it does not even matter whether or not the problems on which we work are important. Each of us experiences these emotions whether he is Einstein or a student who, on his first piece of research, learns the taste of suffering, disappointment and joy.

This knowledge binds us together. We enjoy long scientific talks which would seem to an outsider a torture hard to endure. Even if we work in similar fields we usually have different views, and we may stimulate each other by violent discussions. Every field of research is so specialized that often two mathematicians or two theoretical physicists fail to understand each others' problems and methods. But even then they may feel the bonds created by research though they may gossip mostly about their colleagues, jobs and university life.

There is a level below which our talks seldom sink. I have never heard among scientists the discussion of a frequent topic: "Is science responsible for wars?" We know, perhaps too well, how to avoid glittering generalities. For us Galileo's law is that of a falling stone for which we may substitute in our imagination a simple formula, but never a picture of a bomb dropped from an airplane, carrying destruction and death. To us a knife or a wheel is a great discovery which made the cutting of bread or the transportation of food easy, but we know too well that it is not our responsibility if the same discoveries have been applied to cutting human throats or manufacturing tanks. It is not the knife which kills. It is not even the hand which kills. It is the radiating source of hate which raises the armed hand and makes the tanks roll. We know all that.

The family feeling among us dissipates and vanishes, however, once we leave scientific problems. We have our prejudices, our different social views, our different ethical standards. We are not angels. There are men among us, like Rupp in Germany, who have faked experiments; well-known physicists, like Lenard and Stark, who supported Hitler even before he came to power; mathematicians like Bieberbach, who distinguish between Aryan and Jewish mathematics; and aloof, kind, gentle and progressive men like Einstein, Bohr and Dirac.

Scientists must employ logic, criticism, imagination in their research. As a relief, their brains relax as soon as they leave the domain of science. It is almost as though logic and good reasoning were too precious gifts to be employed outside scientific work.

My generalizations are worth as much as all generalizations of this kind. They are gained by my own experience, from my contacts with scientists, from my own observation. They do not refer to individuals, but I believe they are valid when applied to a majority of scientists.

These scientists are the product of their environment. They have not felt the impact of life. They would like to remain forever on their peaceful island, nursing the belief that no storm can reach their shores. They were brought up in a comfortable feeling of security and hope to retain it by closing their eyes to the struggle of the outside world. They have not strengthened the forces of reaction, but they have not fought them. Indifference has been their sin. They belong to those in Dante's Inferno

> . . . .who have their life pass'd through
> If without infamy yet without praise;
> And here they mingle with that caitiff crew
> Of angels who, though not rebellious, were
> Through neutral selfishness to God untrue.

Slowly, very slowly, through years of bitter experience, some of us have discovered our tragic mistake. We cannot keep our eyes closed. It is not only the problem of the outside world which disturbs our sleep. We can no longer pretend that nothing has happened or that what has happened is not our concern. The storm comes too close to our shores. The waves have washed away many of us and destroyed some of the best laboratories on our island. We look with astonishment at a world which we never wanted to shape, trying to understand the forces of sudden and unforeseen destruction.

The individual is no concern of nature. My story would be irrelevant if it were my story only. But it is not. I belong to the generation of scientists who were forced to view the world outside their island, who had to learn to ask: "What are the forces which try to destroy science? How can we save our kingdom? How can we by our own efforts prevent or delay the decline of the world in which we live?"

We are not fighters; we care little for power; no great political leader has ever arisen from our circle. Not one who has tasted research would exchange it for power. We are trained in too many doubts to employ force and to express unconditional belief. But in the fight against destruction our words and thoughts may count. We shall have to learn the use of words which will be understood, we shall have to sharpen our thoughts on problems which we have ignored before.

The scientist tries to understand the origin of our solar system, the structure of the universe and the laws governing the atom. He has discovered X rays, the radioactive substances, and he has built cyclotrons. He has foreseen the existence of electromagnetic and electronic waves. Out of his thought has grown the technique of our century. But not until today has he begun to notice that the earth on which he moves is covered with sweat and with blood and that in the world in which he lives *"the son of man has nowhere to lay his head."*

The End and
the Beginning

Helen's college year ended. Before going back to Toronto we spent two months in Maine. From our cottage, surrounded by pine trees, we listened to the pounding waves, breathed the peaceful air, talked about the present and past, of the child we expected, as though the air would retain its sweet flavor of peace for the rest of our lives.

News penetrated our isolation. It cried insistently from newspaper stands which we passed on highways, from the excited faces of people carrying papers with screaming headlines. We gave up; we established contact with the outside world by buying newspapers and listening to the radio.

The horizon around our lives grew smaller, narrower with each day. We forgot the pine trees and the ocean. The air lost its freshness. Only the radio remained and increased in size with every hour until it covered our horizon. It brought into our isolation the smell of death, the color of blood, the realization that events to come might be beyond the imagination of any living man.

The butcher came early one morning, put the meat on the table and said:

"The war started today. Poland was bombed."

I shivered and looked at Helen's pale face. We counted the money and Helen put the meat into the icebox.

Somewhere in Poland a mother nurses her child at the same prescribed hour as she did it yesterday and a week ago. But today she does it in the shadow of death.

*Warsaw: German war planes raided the center of Warsaw
this afternoon.*

Go today to your grocer and buy a packet of delicious——

*Hitler in his speech said: "I am putting on my uniform and I
shall take it off only in victory or death."*

All children love its delicious flavor. Try it today——

*This is Paris speaking. The moon shines over the city. Every-
one is quiet. There is no hatred as in 1914.*

Good afternoon, shoppers.

*Polish sources reported there had been seven air-raid alarms
in Warsaw before noon today.*

You will definitely feel cooler——

We decided to drive to New York, collect our belongings
and go to Toronto. As we started back I promised Helen that
I wouldn't read a newspaper during the day of driving. We did
not talk about the war, pretending that nothing had happened.

In the evening we were still in New England. We passed a
tourist home. Helen stopped the car and said:

"Why don't you wait here while I see whether they have
any rooms free."

I nodded. I saw Helen ring the bell. Someone opened the
door and Helen went in. I turned the ignition key of our car
and looked up at the familiar inscription:

Tourist Home. Cosy Rooms with Beauty-Rest Mattresses $1.

Epilogue

EPILOGUE

THE preceding chapters, written in 1940, tell the story of the author's life up to late 1939, when the Germans invaded his native Poland. This epilogue is taken from an article published by the author in the February, 1965 issue of *The Bulletin of the Atomic Scientists*. Professor Infeld died in Warsaw, January 15, 1968. He gave a fuller description of his later life in two small books, originally published in Polish in 1964 and 1968; these have been translated into English by Helen Infeld and are reprinted in *Why I Left Canada: Reflections on Science and Politics* (McGill-Queen's University Press, 1978).

J UST BEFORE the war (when I was forty) I wrote the story of my life. I believed at the time that by doing so I would help counteract the growing anti-Semitism. I wanted to show by my story that anti-Semitism is equally dangerous both for its victims and for those who practice it. I was convinced that my book, *Quest*, would create discussions and influence people. Nothing of this kind happened. My book was almost competely ignored.

Could it have been exhibitionism that led me to write *Quest* and leads me now to write this essay? I do not think so. I want to tell the story of my later years. Why did I leave Canada and return to my native country? . . . Why do I feel it my duty to work for peace? But I must start my story from the beginning, that is, with a brief summary of *Quest*.

If I had listened to my father, who was a good and clever man (about whom I think more and more in my later years), I should not have studied but finished my education at eighteen after four years of trade school. Yet study I did. I passed the University entrance examination just two days before joining the most corrupt and disorderly of armies—the Austro-Hungarian army. It happened some two years before the first world war ended. By bribery or feigning sickness I was able to study at the oldest Polish university, in Cracow. I was a private when the war started and a private when the war ended.

In 1964, the university in Cracow, the Jagiellonian University, became 600 years old. In 1917, before this great and richly celebrated event of Polish learning, there were only one professor of theoretical physics and one professor of mathematics. There was no trace of assistants or seminars. I well remember the lectures on mathematical analysis delivered by the distinguished mathematician S. Zaremba.

There were at most three students present. One nondescript girl who came only occasionally, one very bright student who later committed suicide, and myself. By the time I finished the lectures, I was the only, therefore the best, or the worst, student of the year. The professor in theoretical physics was W. Natanson, whose lectures were technical masterpieces and who kept the students at a distance by the excellence of his delivery and by exaggerated politeness.

I became essentially a self-educated physicist. Such a lack of a school was a blessing for Einstein, but I needed an atmosphere of learning and research more than anything else. When Poland was revived, when new universities were created, theoretical physicists were much needed. The universities at Lwów and Poznan did not have any. When I became the first and only Ph.D., with Professor Natanson as my sponsor, there was not the slightest doubt in my mind that a university career would be open to me if I were not Jewish. But I had to earn a living. The only way open to me was to become a teacher in a small provincial town full of Jews, that had a newly created high school for Jews. The memories of empty life in the towns of Bedzin and Konin will haunt me to my dying days. Those were the years when young scientists do their best work!

I never thought in those depressing years that I would become anything more than a school teacher, and the dream of my life was to move from Polish provincial towns to a university town. Thus I saved some money and went on a job hunt to Warsaw. I was nearing thirty when I found a place as a school teacher in our capital. The work was badly paid. I had to accumulate hours; some years I remember 38 hours of teaching a week. Yet no one gave me encouragement. The professor of theoretical physics in Warsaw never had any research students. My only advantage was a good university library. I was thirty, then thirty-one, then thirty-two. I was aware of having wasted the best and most productive years of my life. I felt a hidden hostility toward me because I was Jewish. Perhaps there was no real hostility. Perhaps it was only my imagination.

I began to write some papers. They were poor papers but they allowed me to escape from the depressing dullness of my surroundings.

If it had not been for Professor S. Loria, an experimental physicist at Lwów University, I would long since be dead among the victims of Treblinka, Majdanek, or Auschwitz. The chair of theoretical physics was vacant for many years in this frontier town. It was Loria's idea to give me an assistantship and let me lecture on theoretical physics. I was then thirty-two years old. Usually such an assistantship is given to a promising graduate student ten years younger than I was; besides, the assistantship was poorly paid. Yet I accepted. From then on the problem of my doing scientific work was solved.

Geniuses do their best scientific work in theoretical physics before they are thirty. Such was the case with Einstein, Heisenberg, Dirac, and many others. Talented theoretical physicists do their best scientific work before they are thirty-five. From then on, the imagination, the most essential factor in theoretical physics, becomes more and more sterile, until at fifty-five or sixty research becomes more a matter of acquired technique than imagination. It is astonishing how few scholars know or would agree with this simple truth. They usually imagine in their steady progress toward the grave that "now" they are doing their best work. There are in the history of science some arguments against my statement (Sommerfield's theory of the conductivity of metals was formulated when he was sixty), but they are extremely rare.

My own scientific work of some consequence began at the age in which the best work of other scientists finishes. In my thirty-fifth year (just before Hitler came to power), I went for a few weeks to Leipzig and saw a flourishing scientific center for the first time in my life. Owing to my meeting Professor van der Waerden, I brought home to Lwów some inspiration for research. Later in Cambridge, I worked successfully with Max Born, whom I regard as my teacher and friend.

The knowledge that I wasted the ten most fruitful years of my life has made of me both a progressive and a good teacher. I am very anxious that war or race hatred be no hindrance to anyone's scientific work. I am anxious that race hatred and war be wiped out from the surface of our earth. I am anxious that my students should not suffer the same fate as I did.

My dream (after my return from Cambridge) was to become a professor in Wilno (a university town on the northeastern frontier now belonging to Lithuania), where the chair of theoretical physics was vacant. By a complicated intrigue, I, who was known as a progressive and a Jew, was prevented from getting it. Thus I was saved again from death in Auschwitz.

I was then thirty-eight years old, definitely on the downgrade. I wrote to Einstein, the noble defender of all persecuted, asking him for help. He offered me a very small stipend in the Institute for Advanced Study in Princeton. I thought then that my "good-by" to Europe was forever. I hoped to find a living on the new continent which welcomed the needy and the tired.

It is time to say something about my political development.

When the Russian revolution broke out I was nineteen years old. In a very vague way I was a liberal. Yet I was rather doubtful about the durability of the revolution. The political situation in newborn Poland appeared to me condemnable and hateful. The only question that seemed to be of interest to the majority of newspapers and Poles was how to get rid of the Jews. The problem of Jewish minorities living in clusters in Polish towns was a constant topic of conversation and unending polemics. This was the number one topic. The number two topic was the danger of Communism. Then the two topics merged into one under the slogan "the danger of Judeo-Communism." The first years of independence were the most dreary.

At that time I was still an undergraduate. I convinced my father to give me some money to go to Berlin for a year, away from the atmosphere of saber-rattling.

I got a passport and went to the town where Einstein, Planck, and Laue were lecturing. Unfortunately, I was not admitted to the university. I stayed there a half year, making use of the splendid library. In Berlin, I met two men, one physicist and one philosopher, and the three of us had weekly meetings devoted to understanding the theory of relativity. One of them was Leo Szilard, who, like myself was young—in his very early twenties. The other was Dr. W., who was older than either of us and through his whole life (up to his death in London a few years ago) a convinced communist. His logical arguments for communism were sincere and seemed to me convincing. Because of my early experiences in Poland they fell on fertile ground. I owe to him my sympathies for the exploited and for many (though not all) aims of the Communist Party. I never joined the party later when I was in the U.S. or in Canada, or even now in Poland. Yet my sympathies are with it, in spite of some revulsions against it (for example, during the Molotov-Ribbentrop pact), and have been deeply rooted in me since the twenty-second year of my life.

In Princeton I worked with Einstein. I described our work together and my relation to Einstein in *Quest*. Einstein belongs to history. Therefore I have no right to hide any knowledge that I have of him. There are, however, some occurrences that I could not have written about when he was alive. Let me tell about them now.

I went to Princeton by way of London in the summer of 1936. There, by sheer chance, I met Professor C. from Princeton. I knew him from my high school days in Cracow. Later he emigrated with his family to Berlin, where he studied very successfully. I had met him in Berlin again. After Hitler, he, being a Jew, had to leave Germany, and he became a professor in Princeton. I was delighted with my good fortune in meeting him. He will be, I thought, my only acquaintance in Princeton. I ask him generally about the U.S., about its universities, and about the probability of getting a job in one of them. He gave me fatherly advice. If I were hunting for a job in the U.S., I had better not work with Einstein.

Yet is was Einstein who had invited me and I thought therefore that I should work with him for at least a while. Einstein was then under sixty—a comparatively young man. I was about twenty years younger. When I accompanied him home the first day we

met, he told me something that I heard from him many times later:
"In Princeton they regard me as an old fool." ("Hier in Princeton
betrachten sie mich als einen alten Trittel.") How much truth
was in this statement?

Before he was thirty-five, Einstein had made the four great
discoveries of his life. In order of increasing importance they are:
the theory of Brownian motion; the theory of the photoelectric
effect; the special theory of relativity; the general theory of relativity. Very few people in the history of science have done half
as much. Einstein was usually very enthusiastic about his current
work and dissatisfied with it some time after it was printed. For
years he looked for a theory which would embrace gravitational,
electromagnetic, and quantum phenomena. I doubted then, and I
still doubt, whether such a unifying theory can be formulated at
all. Yet Einstein pursued it relentlessly through ideas which he
changed repeatedly and down avenues that led nowhere.

The very distinguished professors in Princeton did not understand that Einstein's mistakes were more important than their
correct results. Einstein, during my stay in Princeton, was regarded by most of the professors there more like a historic relic
than as an active scientist.

Einstein suggested that I work on the equations of motion. The
professors in Princeton (remember this was 1936!) were convinced that the work would come to naught. At the beginning I
was as skeptical as the others and wanted to convince Einstein
that the method suggested by him would lead nowhere.

But this turned out to be impossible. To all my arguments,
Einstein replied quietly with deeper counterarguments. In order
to convince him, I had to make more and more calculations. Suddenly I saw the light. Einstein was right. From then on our
collaboration proceeded extremely smoothly.

Yet Professor C. had told me the truth. When Einstein pro-
posed me for a fellowship in Princeton for the year 1937-38 his
proposition was rejected, because no one believed that our work
would bring results. No invitation for a job or lecture came from
anywhere. Princeton was then full of younger and better scholars
working on more attractive subjects. I still—so they thought and

said—could return to Poland. Then Einstein and I decided to write the *Evolution of Physics*. Thus for the third time, I was saved from death in Auschwitz, for the second time by Einstein.

After we wrote our big paper on the equations of motion (Einstein, Infeld, and Hoffman), and after the *Evolution of Physics* appeared in 1938, my situation changed completely. The offer that I received and accepted was from Professor Synge for a lectureship at Toronto University. As my friend Professor H. P. Robertson had told me, I had to start again from the bottom of the academic ladder. At forty, I believe I was the oldest lecturer in Toronto. However, I climbed the academic ladder quickly enough, becoming a full professor in a few years.

In 1939, Poland's tragedy started the world war. For the second time in their lives my generation experienced the horrors of war—I mostly through fear of what was happening to all those whom I loved and who remained in Poland. Like many scientists, I did what I could to help the war effort. I worked with my colleagues in Toronto—first on ballistics, then on radar. Since I was a theoretician and not an atomic scientist, my field was rather restricted.

I was well aware that the future of the entire world was at stake. Only after Stalingrad could I breathe more freely. I remember still my many dreams in which I was beaten and tortured together with my friends and members of my family.

My fears were well founded. My younger sister, who was also the best friend I ever had, vanished suddenly in Cracow and no one knows how she died. Her husband was shot on the street and their young son died of illness and hunger. My older sister died in Bergen-Belsen . . . but I had better stop.

In 1945, the war ended and the government of the People's Republic of Poland was recognized by almost all nations. By then, I was a Canadian citizen. I wanted to see Poland, but I was afraid of seeing it. I was afraid that the country would seem to me a great cemetery where the members of my family, among three million Jews, were buried. Yet at the same time, I felt a strong attraction to my native country, which had abandoned its half-feudal past and was building a new socialistic order. Thus, when an invitation came from the Ministry of Higher Education in

Warsaw to come to Poland for a short visit, I accepted it gladly. This happened in late April 1949.

If I were asked what are the two most characteristic features of Canada I would say "decency and dullness." At least so it was up to 1950. The weekdays were bad enough but the Toronto Sundays, with closed restaurants and movies, but with open churches, were depressing in the extreme. I prayed to God that if I died in Toronto He would be good enough to take care that this last event of my life should happen on a Saturday to save me from one Toronto Sunday.

During the war, the atmosphere in Canada, though far from the centers of fighting, was sad and depressing. The heavy clouds under which we all lived began to lift only when Germany was beaten.

In 1945, during an afternoon visit to the great Polish poet Julian Tuwin (who also was in Toronto at that time), I learned from him about the atomic bomb and Hiroshima. He had just heard the news on the radio. I was astonished by the news, and shaken. I had heard gossip (mostly through my students who were a part of the Montreal group) that some work on this problem was going on, but I didn't know that the work was near completion or finished. Later, I read Smyth's book, which gave the details.

Billions of words were then written about this terrible event. They can be summarized in a few wrong sentences: "We Americans are the only ones in the world who have the know-how of the bomb. As long as we keep the great secret we are safe. Let's beware of spies and this century will become an American century."

The smugness of some officials had no limits. Thus, General Groves was quoted as authoritative when he said the Russians would have the atomic bomb in twenty-five years at the very earliest, or perhaps never.

From 1945 to 1949 I gave some hundred popular lectures on the elementary physics of the atomic bomb and the importance of preserving peace. My message was very simple: in three years the Soviets would also know how to construct the A-bomb. Its so-called secret did not exist. This short time should be used to work for mutual understanding.

On the initiative of the Canadian Institute of International Relations (sponsored, I believe, by Carnegie funds) I travelled with my simple peace message across Canada, meeting new people in clubs and churches. The discussion gave me a good idea of the damage done by the press to the understanding of the problems of peace. For example: after I had stressed as forcibly as I could that there was no secret of the atomic bomb, I often was asked from the floor: "How can we prevent Russia from taking our secret from us?" Very few people (if any) believed that the Soviet Union would soon be able to manufacture the A-bomb. I predicted it almost to the month. I was very proud that my prediction came true and did not foresee at that time how it would lead to bitterness.

My journey to Europe in 1949 was the first since I had come to the North American continent. I gave lectures in Dublin, where my old friend J. L. Synge (who had left Toronto quite a few years before) was then in Manchester and Birmingham. But my chief interest was Poland. There I spent almost three weeks. I visited the ruined cities of Warsaw and Wróclaw, the royal city of Cracow which looked to me as beautiful as ever, yet stranger than ever. Its Jewish section of Kazimierz still existed but there were no Jews left in it.

In Warsaw I found the only woman friend of my youth who had lived through the horrors of occupation, whom I saw daily and who taught me about the realities of present-day Poland.

In the evenings, walking through the ruins of a destroyed, dark city, I had the impression that I had found myself suddenly on the moon. During the day, I often took part in meetings where people without any sense of humor used the Marxist phraseology with great earnestness. In private, many of the same people liked to talk against the government, and everything Western glittered to them with elegance and beauty.

Poland in 1949 seemed poor in the extreme if compared to Canada. Yet this poverty was not repulsive to me; on the contrary, I felt annoyed by the memory of the smugness and richness of the West. To this country, whose difficulties I saw more clearly every day, I became progressively more attached. I was convinced that if I stayed here longer I could be of some service to Poland. What contrasted most with my experiences in Poland was the frustration I felt in Toronto, much heightened after Synge's departure. All my efforts to help build a reasonably good department of theoretical physics had turned to nil. The good young people left Toronto after receiving much better offers from the U.S. Because of the stinginess of the administration I could not get anyone worthwhile. How different life in Poland seemed to me! It was full of a dynamism which seemed to me to be its most essential feature.

The Polish Minister of Higher Education suggested that I come to Warsaw and work there permanently. I told him how I felt about Poland: like a husband who is in love with a wife older than himself, limping, and blind in one eye. But, I said, I have an American wife and two young children, used to the great comforts offered by the West. I can try to transplant them only for one academic year at most. I am fifty-one, I said, and it would not be easy for me to make the great change either. After some discussion we agreed on the academic year 1950-51, providing that my wife agreed to it. The minister assured me of a minimum of comfort, which was the most Poland could offer me: a four-room apartment, a car at my disposal, and a telephone (then a very rare thing). I was sure that after an academic year in Poland and, in general, in Europe, I would be cured of my attachment to Poland and be glad to return to Canada.

If I think about my past life and if someone were to ask about my most unhappy time, I would not mention the years I spent in Konin or Bedzin, I would not even mention the time Halina, my Polish wife, died, for my life then had one virtue: I was young. Without much hesitation, I would mention the first half of the year 1950.

I came back from Poland to Canada to face the academic year 1949-50. The academic year in Toronto is very short. Lectures end around the middle of April. Therefore my short visit to Poland was still during the Warsaw academic year. The year in

Toronto promised to be especially unpleasant. My best student and colleague left Toronto for the U.S., where he had been offered a much higher position than in Toronto. I felt scientifically more lonely than at any time before. The cold war was intensifying. No one asked me about my impressions of Poland. Canada with its wealth and my colleagues with their smugness got on my nerves. They knew all the answers and the Communists were never right, but always to blame.

I told the head of my department (who was also the dean) that I intended to ask for a leave of absence for the next year. This was my first such request after twelve years of service. I told him that I wished to go to Europe and spend a large part of my leave in my native country. He agreed to it without much enthusiasm and asked me to write a formal application. Some time later, he told me that the president of the university had no objection to my plan. Einstein's attitude to my plan was also encouraging.

My wife, who had never been to Europe before, looked forward to this trip. Thus I thought that everything was arranged for the next year and I settled down more calmly to my daily teaching routine. I wrote to the Polish Ministry of Higher Education that I accepted their offer for the academic year 1950-51. My wife remembers well that I told her, "It would be difficult for you and our children to go for longer than a year to Poland, but you will enjoy one year in spite of the privations which are awaiting you there."

One of my many friends in Toronto was Professor R. One day he told me that a journalist wanted an interview with me. This was not very unusual. Because of my writings and lectures, I was fairly well known in Canada (to some, as a dangerous progressive) and such a request was not isolated. I only hold against Professor R. that he did not tell me what kind of journalist this one was.

This newspaperman came to my home and introduced himself as the new editor of *Ensign*. I did not know anything about this weekly. Later I learned that it was a clerical-fascist one, issued in Montreal and sold almost exclusively in Catholic churches.

I did not like the man. When he asked me about my future

plans I answered him briefly that I intended to spend some time in Poland and then return to Canada. I gave him three issues of the *Scientific American* in which my report about my journey to Europe had appeared.

A few days later, a news agency (I don't remember which) rang me up from Montreal. I was asked: "Do you know what is appearing about you in the next issue of *Ensign?*" I answered: "No. Would you like to read it to me?" A monotonous voice began to read the long article. As I found out later, almost the entire issue was taken up by my story. The arguments were simple enough and could be summarized in these points: (1) I was a collaborator of Einstein, and as such I possessed the secret of the atomic bomb; (2) additional proof: I had predicted correctly when the Russians would have the bomb; (3) I wished to go with my secrets behind the iron curtain.

"Do you wish to make a statement?" asked the voice on the telephone. I tried to be calm. I said that it was all nonsense, that I had worked with Einstein on relativity theory and that my knowledge of atomic phenomena was what could be found in textbooks, that I intended to spend some time in Poland and return to Canada after my leave of absence.

I did not know at the time that the *Ensign* article would be quoted in all the Canadian dailies. An insignificant periodical became nationally known through this publicity. My denials were also printed, but to the majority of simple Canadians the accusation probably seemed just. Only a traitor could wish to cross the iron curtain. The only satisfactory denial would be that I never intended to go to Poland.

Interviewed on this subject, the president of the university claimed that it was not he who had decided about my leave of absence for the next year; the granting of leaves belonged to the Board of Governors. He was right formally, but in reality the Board of Governors rubber-stamped the decisions of the president.

My visit to Poland was endangered. I hoped that perhaps the news about my leave to Poland would die down and people would forget about the silly and vicious article in a parochial weekly.

There is a rule or custom in the Canadian Parliament that a member is allowed to bring immediately to the notice of the Parliament events which endanger the state.

At that time, of the two parties that ruled Canada—the Liberals and the Progressive-Conservatives—the first was in power. The leader of the opposition was George Drew, a handsome, middle-aged man generally called "Gorgeous George."

On an evening of early March 1950, I was called by a news-paperman, who told me that half an hour ago Drew had brought up my going to Poland in Parliament as a matter of extreme urgency. He quoted *Ensign* liberally, called me an atomic scientist who had many secrets and would peddle them behind the iron curtain. I do not think that Drew was so stupid as to believe all this rubbish, but he thought that anything that would embarrass the government was good enough.

Then all hell broke loose. The next day in Toronto's only morning paper, the *Globe and Mail*, a long article with my picture appeared on the first page. I felt I was being called a traitor by the country of my adoption. I waited for a reply from the government. None came. I remember well the awful day at the university. On my way, I saw newspapers lying on the wet pavement, people treading on my picture, looking at me, and recognizing me as the "traitor" to Canada. That day, the bank clerk gave me my money without any conversation about the weather. The students were especially quiet, pretending that nothing had happened.

Yet I found out a little later, through editorials, that at least a great part of the Canadian press had more sense than to believe in the accusation. But all the press and the majority of my friends were convinced that I should not go to Poland. The president of the university offered to double my salary if I didn't go. All these pressures had just the opposite effect on me. I began to hate my life in Canada. I began increasingly to long for Poland, though I knew then that if I went to Poland I might lose my job in Canada. Thus, I was confronted with two alternatives: either to remain in Canada, regarded as a potential traitor, keep quiet, conform to the majority, and cease any work for peace, or to spend the rest of my life in Poland, in much less physical comfort and in a poorer scientific atmosphere.

My wife thought and decided exactly as I did. We chose Poland. I told my friends that I intended to spend my vacation in England to think over my problems away from the pressures of Canada. But really I had decided to go straight from England to Poland. My family was to join me later through New York on the Polish MS Batory.

Later I found out what my family had to experience during my absence: incessant anonymous telephone calls. Once, when my seven-year-old daughter was not home, my wife received a faked telephone call—a little girl's voice cried, "Mummy, I'm hurt!" Then there was also the steady vigilance of the Royal Canadian Mounted Police. Or a call purporting to be from a flower shop (when my wife was not at home) asking, "Could you give me the name of the ship on which Mrs. Infeld will sail since we have an order to send her flowers?"

Only when they were in New York before departing, did my wife have some peace and relaxation. My family joined me in Poland in July 1950. My two children knew nothing about our plan. They were too small to keep the secret.

Let me first finish the Canadian story. While in Poland I received via London a letter from my dean in Toronto. I was told that I must either come back to Toronto for the beginning of the academic year or cease to be a professor at Toronto University.

I answered the dean with a letter of resignation in which the entire story was once more formulated, of course from my own point of view. For a few days I became famous. In Canada, headlines and long articles were written about the case of a scientist vanishing behind the Iron Curtain. I believe there was no important paper in Canada, the U.S., England, or Australia which did not carry the great news. Of course, the Canadian articles were the longest and many people were interviewed on this issue. The vice president of the Research Council of Canada assured the public that I could not have taken any atomic secrets because I did not have any. Why had he not made this statement a few months before?

After a few days the entire fuss died down. While in Poland, I was asked by the minister of Canada (only later was the post raised to an embassy) to give back my Canadian passport. This I did, returning to my original Polish citizenship. Some years later, my children unexpectedly received letters from the Canadian embassy notifying them that by special Order in Council their native Canadian citizenship had been revoked—I believe illegally.

Since then, I never go officially to the Canadian embassy, though I was friendly with the last Canadian ambassador—an exceptionally intelligent and charming man. (He was helpful in returning the Polish treasure, saved from war destruction and later hidden by the Duplessis fascist-type administration of the province of Quebec.)

There is one more interesting occurrence in connection with Canada. Not long ago (in 1962), one of my friends was invited to a dinner at the Canadian embassy in honor of George Drew. The once Gorgeous George had grown old, and lost his political ambition. He must also have mellowed in the meantime. He asked with great solicitude after my health and begged my friend to give me his best wishes.

The Polish government and the party has done everything they could to further the development of physics in Poland. They have built an institute for me and I never had any difficulty in appointing new men to my department. When I came to Poland at the age of fifty-two, there were two other professors. One was about sixty, the other about seventy. There were no Ph.D.'s and no graduate students in the entire department of theoretical physics.

Today, I have a big department, with me as its eldest member and, unfortunately, the next oldest professor is only thirty-eight years old. The gap is due to the war, which wiped out the generation between us. We have in our department of theoretical physics six associate professors (docents)—it is a big department which is still growing.